Reclaiming Contaminated Land

Reclaiming Contaminated Land

Edited by
T. CAIRNEY, BSc, PhD, CEng, MICE, MIWES

Dean of Construction
Liverpool Polytechnic

Blackie
Glasgow and London

Blackie & Son Limited,
Bishopbriggs, Glasgow G64 2NZ

7 Leicester Place, London WC2H 7BP

First published 1987

British Library Cataloguing in Publication Data

Reclaiming contaminated land.
1. Reclamation of land 2. Soil pollution
I. Cairney, Thomas
627'.5 S605
ISBN 0-216-91874-X

Photosetting by Thomson Press (I) Ltd, New Delhi
Printed in Great Britain by Bell & Bain (Glasgow) Ltd

Preface

Chemically contaminated land has only recently been recognized as an immediate or long-term potential hazard, and published guidance on how to tackle such land has been sparse. Indeed much of the available technical work emphasizes the risks and dangers, rather than indicating safe and economical strategies of reclamation.

This book provides a comprehensive treatment of the important aspects of land reclamation. Its basic aim is to dispel the myths that have become associated with the subject and to indicate methods and strategies that can be used for safe and economical reclamation. The authors concentrate on the more important facets of reclamation and indicate where advice and information is more or less certain. As in any newly developed field there are still uncertainties and, for this reason, not all the chapters contain equivalent amounts of detail.

All the authors have expertise in the field of land reclamation, and differences in emphasis between authors reflect the present state of the subject. Overall, the book emphasizes that contaminated land can be reclaimed safely provided that sufficient attention is paid to detail and that the proposed end use of the land is appropriate and based on a detailed knowledge of the site.

Interest in the subject of land reclamation is expected to increase, not only because of the large number of known sites but because of the economic advantage of re-using urban land rather than green-belt or productive agricultural land. As more organizations become involved in this work, our experience and effectiveness will grow, and present uncertainties will be dispelled. Contaminated land reclamation may then have become another routine recycling of an essential resource.

TC

Contents

Contributors

D.L. Barry, BE, CEng, MICE, MIHT
Senior Environmental Consultant, W.S. Atkins & Partners, Epsom

R.M. Bell, BSc, PhD
Deputy Director, Environmental Advisory Unit, University of Liverpool

T. Cairney, BSc, PhD, CEng, MICE, MIWES
Dean of Construction, Liverpool Polytechnic

R.C. Haines, BSc, PhD
Director, Ecotec Research and Consultancy Ltd, Birmingham

Mary R. Harris, BSc, PhD
Research Assistant, Ecotec Research and Consultancy Ltd, Birmingham

D.W. Lord, MChemA, CChem, MRSC
Public Analyst and Agricultural Analyst for the Counties of Lancashire and Cumbria

G.D.R. Parry, BSc, PhD, ALI
Director, Environmental Advisory Unit, University of Liverpool

M.A. Smith, BSc, CChem, MRSC, MICeram
Associate Director, Bostock, Hill & Rigby Ltd, Birmingham

1 Recognition of the problem

MARY R. HARRIS

1.1 Introduction

1.1.1 *What is contaminated land?*

Land can contain substances which are undesirable or even hazardous as a result of natural phenomena such as the deposition of materials due to flooding or landslides, the eruption of volcanoes or the mineralization processes which determine the natural composition of soils. Equally, human activity, in terms of the use of resources and the discarding of wastes, can result in slight or profound disturbance to the soil environment. As a direct consequence of the nature of contemporary industrial activity, in combination with the formulation of land use policy in modern industrial societies, there are occasions when the presence of certain substances on land either requires active control and management of the land and the substances contained therein, or places serious limitations on the future use of the land.

Under certain circumstances, and according to some definitions of contaminated land, the simple presence of hazardous substances will be sufficient reason for action to be taken to ensure that potential hazards are minimized or even eliminated. However, in the UK the emergence of contaminated land as a matter of concern stems more from the gradual recognition that some land, as a result of technological change or indiscriminate disposal, has become 'spoilt' in such a way that it is no longer immediately suitable for use in a different capacity. However, policies over recent years aimed at protecting greenbelt and agricultural land by limiting development of them, have meant that former industrial land or land with a complex history of former use is frequently the only development land available. Much of the work which has been done in relation to contaminated land has been aimed at overcoming the technical difficulties, minimizing risks and reducing the high costs associated with redevelopment in order to recover valuable land resources while at the same time maintaining development restrictions on better quality land.

While the study and management of contaminated land is a relatively new subject, it is nevertheless possible to identify the main characteristics of contaminated land. It is found typically, but not exclusively, in urban or industrial areas. It is usually land which has been used for industrial or waste disposal purposes and as a result contains substances which may have detrimental effects on both living organisms and abiotic materials. It is this last characteristic that distinguishes contaminated land from other categories of disused or derelict land. There may be physical dangers associated with the latter (cf. the Aberfan disaster of 1966), an example of the consequences of uncontrolled 'spoiling' of land on a massive scale (1), or there may be other risks involved with the re-use of derelict land, such as the potential damage to buildings erected on land with poor ground stability. However, it is the hazard arising directly from the presence of contaminants which causes the greatest concern with respect to contaminated land and, in the UK, the hazard is usually considered to exist in relation to the redevelopment of the land.

A typical general description of the circumstances under which contaminated land may arise and the implications which exist in relation to redevelopment is as follows:

> 'The use of land for industrial purposes or for waste disposal may result in chemical contamination which can restrict or prevent subsequent redevelopment because of immediate or long-term hazards to human health (directly or indirectly), to plants, to amenity, to construction operations, or to any buildings and services' (2).

Various workers have striven to provide a formal definition of contaminated land. One of the most common is:

> 'Land that contains substances that, when present in sufficient quantities or concentrations are likely to cause harm, directly or indirectly to man, to the environment, or on occasions to other targets'

A recently completed international study on contaminated land adopted this definition (3), while the British Standards Institution (BSI) draft Code of Practice on the identification and investigation of contaminated land (4) used a similar working definition:

> Land that contains any substance that when present in sufficient concentration or amount presents a hazard. The hazard may
>
> (a) be associated with the present status of the land
> (b) limit the future use of the land
> (c) require the land to be specially treated before particular use.

Definitions of this type have recently been challenged by the Department of the Environment (DoE) (5), and a new operational definition of contaminated land proposed:

> (i) Land which because of its former uses now contains substances that give rise to the principal hazards likely to affect the proposed form of development, and which
> (ii) Requires an assessment to decide whether the chosen development may proceed safely or whether it requires some form of remedial action, which may include changing the layout or the form of the development.

It will be seen that the exact format of these various, apparently similar, definitions not only has repercussions for the estimation of the scale of the

Table 1.1 List of 'contaminating' industries*

Gas works and similar sites	Pharmaceutical industries
Scrapyards	Iron and steelworks
Railway land	Tanning and associated trades
Sewage works and farms	Wood preserving
Mining and extractive industries	Dockyards
Waste disposal	Chemicals
Metal smelting and refining	Oil refining and storage
Metal treatment and finishing	Explosives industry
Paints and graphics	Asbestos manufacture and use

*Based on references (2, 3).

problem but also provides some insight into how various authorities regard the subject of contaminated land.

1.1.2 *Where does it occur?*

Contaminated land may arise in a number of ways, many of which result from manufacturing and other industrial operations. These pathways include losses of materials from industrial plant, for example through fugitive or stack emissions; accidental spillages and leaks; the dumping of wastes (both on and off industrial premises); the discharge of effluents and the storage of raw and waste materials. Under some circumstances, substances originally released into sewers can be returned to the land in the form of contaminated sludges and sewage which have been applied for agricultural purposes.

Particular types of industrial, and other activity, appear to be frequently associated with land contamination and Table 1.1 shows a list of such activities. It should be noted however, that some authors consider as suspect all former industrial uses of land (3). In some cases only a single, previous use of a site will be identified and a characteristic pattern of contamination will be observed. However many factors, such as the age, capacity and type of installation or process will serve to reduce the probability of a characteristic pattern of contamination being present. In other cases a number of former uses will have contributed to the contamination present (6). Under these latter circumstances, a simple evaluation of the immediate past use of a site will not be sufficient to indicate the true condition of the land. The different types of contaminated land which may be encountered are discussed further in Chapter 2.

1.1.3 *How common is it?*

Although there is considerable practical experience of the investigation and restoration of contaminated land, there is relatively little information on the amount of land in the UK which may be affected by contamination. In part this is due to the fact that only a relatively few sites have been well documented and there is no national mechanism, such as a land registry specific to contaminated land, by which the extent of the problem may be judged, although this is one area which the Royal Commission on Environmental Pollution suggests requires improvement (see section 2.1). There are also difficulties in relation to the choice of definition of contaminated land. The new operational definition suggested by DoE is drawn strictly within the context of redevelopment and clearly infers that land which is contaminated but still in active use must be excluded from the definition. In contrast, a definition such as the one adopted by the BSI draft Code of Practice states that the hazard presented by contaminated land may be associated with the present status of the land. Therefore, land which is contaminated to the extent that it poses a hazard, but which is still in active use, could legitimately be included within the BSI definition. Clearly any estimate of the size of the problem would be highly dependent on the choice of definition. This fact is acknowledged by DoE to the extent that when offering an estimate of 'at least 10 000 hectares in England alone' of land which may be chemically contaminated and currently available for reclamation or redevelopment, reference is made to the fact that a much higher figure would be derived if land which is in active use or contaminated as a result of natural processes were to be included (7).

The current lack of information on the number, size and location of contaminated sites severely disadvantages Local Authorities in relation to their statutory obligations to produce local and structure plans. Local Authorities are unlikely to make the best possible decisions on redevelopment matters if the existence of contaminated sites within their boundaries goes largely unnoticed. Furthermore, the cost and time penalties of failing to take account of potentially contaminated sites at some stage prior to redevelopment places considerable strain on the financial and staff resources of Local Authorities and developers alike. It has been suggested that the use of 'prospective surveys' can do much to avoid these difficulties. The aim of such surveys is to identify potentially contaminated sites on the basis of their former use in order to alert Local Authorities to the presence of suspect sites before comprehensive development plans have been made and to provide information in relation to specific development proposals.

Prospective survey methods generally rely on the use of a range of information sources including maps, commercial directories, land use survey data, rating records and deeds, aerial photography, Local Government records and local knowledge to ascertain the previous use of a site or area. Since certain activities are likely to result in the formation of contaminated land it is possible to decide to what extent a site or area will present problems in relation to particular forms of development. The usefulness and limitations of the various information sources have been described by Barry (8) and JURUE (9) and reference should also be made to the work of Dumbleton (10, 11). Some records are limited by the infrequency of publication or insufficiency of detail or are too costly to research to provide all the information required although a combination of the use of all such sources will be adequate. An example of the way in which prospective surveys can be used to supply information on the amount and location of suspect sites is provided by a recently completed survey of potentially contaminated land in Wales (1). The survey, limited to sites > 0.5 hectares in size, identified over 700 sites, representing an area of 3,787 hectares, in which the soil is believed to be contaminated.

While providing useful information, prospective surveys are capable only of identifying potentially contaminated land and are not designed to confirm the presence of contamination. This limitation has been recognized and an alternative procedure has been suggested which, while providing detailed information on contaminant types and concentrations, manages to avoid the high cost of intensive, site-by-site investigations (12). Surveillance techniques combine the use of maps and land use data with laboratory analysis of soil samples taken from representative locations within the area under consideration. The result is an area-based spatial prediction model from which maps of soil contamination can be obtained. The technique has been used to assess background levels of metals (including lead and cadmium) in the soils of the West Midlands Metropolitan County. A similar technique was used to provide a Borough Authority with information on which to base development decisions. Soil contamination maps produced as a result of the survey, when compared to the West Midlands Structure Plan, indicated that while much of housing development was to take place in relatively uncontaminated parts of the Borough, two were designated for sites showing metal levels in excess of the recommended Government guidelines for residential development. The survey not only enabled the Authority to consider making a more detailed assessment of the sites involved but also caused it to look again at other sites which had been designated for short-term allotment use.

Clearly, surveillance techniques provide quantitative evidence of soil contamination which prospective surveys are not able, or designed, to achieve. However, one limitation associated with surveillance techniques concerns the range of contaminants which can be included in the survey on the grounds of cost.

In spite of the limitations attached to prospective surveys and surveillance techniques, these procedures are clearly of practical value to Local Authorities and others with whom responsibility for development schemes rests. In addition these methods do appear to have some role to play in providing information on the scale and distribution of suspect or actually contaminated land—information which is conspicuously failing to materialize from other sources.

1.1.4 *How is it managed?*

It is clear that while it is possible to give some information on the general properties of contaminated land there remains some difficulty in obtaining a universally acceptable definition which not only embodies the known qualities of contaminated land but also, in the UK context, imparts some message as to how it might be safely redeveloped. The main reason why contaminated land, to some extent, defies precise definition is that it is fundamentally and inherently complex. It will be seen that the complexity of contaminated land is manifest at all levels of management, from the point of identification through investigation and subsequent redevelopment.

The diverse origins of contaminated land have already been described (see section 1.2 and also Chapter 2). The situation is complicated by the fact that some land, particularly in densely populated areas, has a history of multifunctional use. This presents difficulties for both identification and investigation as well as for the development stage.

Contaminated land also displays an alarming variety in terms of its chemical and physical characteristics. Sites vary with respect to the nature of the contamination in relation to the chemical species present, their spatial distribution and the concentrations recorded. The types of contaminants which may be encountered in practice include some metals and their compounds; oils and tars; other organic constituents; cyanides and sulphates; asbestos; toxic, flammable and asphyxiant gases; combustible materials such as coal dusts and domestic wastes, and, although beyond the scope of this book, radioactive materials and biologically active residues and agents. The physical state of the site may add to these difficulties. In addition to redundant buildings above ground there may be present large buried structures and vessels which may act as loci of high concentrations of contaminating substances.

The potential hazards associated with contaminated sites are correspondingly varied. There may be short-term hazards, such as the risk of direct contact with corrosive substances, the inhalation of toxic dusts and fumes or the immediate risk of fire and explosion. In some cases, fire and explosion hazards may not be immediately apparent and may emerge over the long-term perhaps due to inadvertent containment of active materials during redevelopment (see Chapter 11 for information on gas generating sites). Other long-term potential hazards include the ingestion of contaminated food or water, and regular and intensive contact with contaminated substrates, particularly by small children. It is possible to identify 'target' groups in respect of potential hazards and again the range of critical groups reflects the diversity of the hazards themselves. The target group may be individuals engaged in site investigation work, construction workers involved in development operations or the target group may be the eventual users or occupiers of the site who are at risk by virtue of prolonged and close proximity to the site in daily life. Some authors would also seek to protect a wider population which may be exposed to hazards by way of wind- or waterborne contaminants arising from a site. Plants and animals, both terrestrial and aquatic, may require some protection, while under some conditions, the inert materials used in buildings and services may constitute a target group. For example, concrete is known to be adversely affected by high levels of sulphate in soils while some organic compounds can attack plastic coatings, seals or connectors used in underground services, or pass through plastic pipes to taint water supplies. The reader is directed to Chapter 3 for a more detailed account of the range of potential hazards and critical groups which may be cause for concern in connection with the development of various types of contaminated land.

Given that there is considerable potential variation both within, and between, contaminated sites and the importance of protecting the welfare of those people and materials in contact with them, the fundamental role of a thorough site investigation is obvious. The purpose of site investigation is to establish the nature and extent of the contamination in order to decide the best course of action. Much of the success and operational efficiency of the redevelopment stage, whether it includes the use of remedial measures or not, depends on a well planned, well executed and productive site investigation. While the complexity of contaminated sites makes it extremely difficult to formulate a 'standard' site investigation procedure which will be suitable in all cases certain general principles should be observed and the reader is directed to Chapter 4 which covers appropriate site investigation strategies.

While there should not be an automatic assumption that extensive remedial or reclamation methods must be an integral part of all redevelopment plans, inevitably there are situations in which action to ameliorate conditions on a site are considered to be desirable or even essential. As is the case with all other aspects of contaminated land management there is no standard reclamation technique which will be appropriate in all possible circumstances. Instead there is a requirement to match the type of remedial action to the nature of the hazards and the needs of the proposed development. For detailed information on currently available methods of reclamation, and related aspects of safety, finance and landscaping the reader is directed to the final chapters of this book.

1.2 National response to contaminated land in the UK

1.2.1 *Central government response*

During the 1970s certain Local Authorities began to experience problems in relation to the redevelopment of particular sites which were contaminated as a result of their former use. In one case, that of the Greater London Council's new-town development at Thamesmead on the site of the former Woolwich Arsenal, development on part of the site had already begun when severe contamination, associated with former munitions manufacture, town gas generation and the dumping of waste materials, was encountered (13). Another large scale development at Beaumont Leys in the Midlands ran into problems when high levels of metals, including lead, cadmium and zinc were discovered in the soil. Here, contamination had resulted from the use of the site, over several decades, as a sewage farm (14). In both cases large scale residential use of the sites had been planned and it was a combination of the size and nature of the development which prompted the respective Local Authorities to express their concern.

Inquiries were made to two Departments, the Department of the Environment (DoE) and the Department of Health and Social Security (DHSS) regarding the health risks associated with the development of these sites. In responding to these inquiries and forseeing that pressure to develop old industrial areas for housing, schools or amenity use would increase, it was decided to establish a Government Committee to co-ordinate advice from various Departments and make such advice available to Local Authorities. In 1976, the Interdepartmental Committee on the Redevelopment of Contaminated Land (ICRCL) was set up with representatives from DoE, DHSS, the Welsh Office, the Health and Safety Executive and the Ministry of Agriculture, Fisheries and Food. The Scottish Development Department is now also represented on the Committee. While the principal concerns of the Committee were considered to rest with issues of health and safety, and effects on the environment in general, it was recognized that the redevelopment of contaminated land would have implications for building and construction matters and for this reason the DoE membership of ICRCL included representatives from the Housing and Planning Directorates, the Property Services Agency and the Building Research Establishment.

The terms of reference originally developed for ICRCL are described in a progress report on Committee activities published in 1979 (15). These included:

> 'to develop and co-ordinate advice and guidance on human health hazards arising in connection with the re-use of contaminated land, to develop and co-ordinate advice regarding possible remedial measures; to make such advice available to Local Authorities; to advise the Department of the Environment as to the appropriateness of the methods selected by Local Authorities; and to identify research needs.'

The aims of the Committee were later amended to include reference to the effects of land contamination on building materials.

In order to satisfy these terms the Committee embarked on four main programmes of work. The first was to establish an advisory service to deal with inquiries on the redevelopment of contaminated land. Initially the service was geared to cope with pressing needs of immediate concern; however it was the intention of the Committee, as information and experience were gained, to develop a comprehensive system of guidance which would both support and encourage an active programme of development. While Local Authorities were to be the main recipients of such advice, information and assistance were also, on occasion, forwarded to the private sector and in fact a two-way communication system came into operation. Problem sites were referred to ICRCL from a variety of sources including Local Government Departments, regional DoE offices and private consultants while the Committee itself sometimes approached Local Authorities directly to learn of their experiences in connection with contaminated land.

The Committee regarded the publication of information to be a second important function and over the lifetime of the Committee a number of guidance papers of both a general and specific nature have been published (16–21).

A third important role of the ICRCL was considered to be one of initiating and supporting research into those aspects of contaminated land development where little information was available. To this end, the Committee, through DoE, set up a number of research contracts some of which were aimed at obtaining a better understanding of contaminant behaviour under various types of remedial treatment. Full descriptions of such research and its findings can be found in a number of published accounts; on the effectiveness of remedial treatments in relation to metal uptake by plants (22–27); engineering methods of reducing metal levels in soils (14); problems of contaminant migration (28–31) and the performance of past reclamation work (32). Research programmes in a number of related fields have also been carried out including the combustibility of fill materials (33) and the problems presented by methane (see Chapter 11), the redevelopment of former gasworks sites (34); prospective survey methods (8.9) and a review of in-situ treatment options for use on contaminated land (35).

The fourth direction in which ICRCL spent its energies was to bring the subject of contaminated land into open debate. ICRCL was involved in the organization, in 1978, of a Society of Chemical Industry conference on the reclamation of contaminated land (36). The conference proceedings, now nearly six years old, remain a valuable reference source.

Since its inception in 1976, ICRCL has continued to develop guidance for Local Authorities and others with interests in contaminated land. There has also been Departmental involvement in other programmes of work of direct relevance to the role of ICRCL as a central source of advice. Two important examples are the NATO/CCMS (Committee on Challenges to Modern Society) pilot study on contaminated land (3) and the preparation of the BSI draft Code of Practice on the identification and investigation of contaminated land (4).

The NATO/CCMS study group commenced work in November 1981 with the main aim of reviewing current knowledge of remedial measures particularly in relation to their long-term performance. The study, in which seven member countries (Canada, Denmark, Federal Republic of Germany, France, Netherlands, UK and USA) were represented, involved individual participating countries taking a lead in one or more study areas, of direct national significance (see Table 1.2), and exchanging information in a number of other areas. The study served two useful purposes. Firstly, by drawing on individual national experience, the study derived information of value to all

Table 1.2 The main NATO/CCMS study areas*

Study area	*Lead country*
In-situ treatment of contaminated sites	USA
On-site processing of contaminated soil	Netherlands
Covering and barrier systems	UK
Control and treatment of groundwater	Canada
Rapid on-site methods of chemical analysis	USA
Long-term effectiveness of remedial measures	FRG
Toxic and flammable gases	USA

*Based on reference (3)

member countries and to the international community at large, and secondly the study provided a useful technical complement to the work of the OECD Waste Management Study Group on administrative and legal issues (37). The final report of the Group, containing details of the findings, conclusions and recommendations, was published in 1985 (38).

DoE is also represented on BSI Technical Committee EPC/47—Land Quality which is currently preparing guidance on the identification and investigation of contaminated land. The draft Code of Practice document was issued for public comment in December 1983, and the advice is intended to complement existing BSI guidance, (BS 5930: 1981) on the assessment of sites in relation to building and civil engineering work, by offering detailed information on sites where man-made contamination is a major feature. Summary accounts and critical appraisals of the draft Code are available (39,40) and for further information the reader is directed to Chapter 4.

The role of ICRCL as a central source of informed advice is an important one. The policies developed by the Committee can do much to influence the opinions on, and attitudes to, contaminated land particularly those held by relatively inexperienced organizations who are encountering problems for the first time. Undoubtedly, the quality of the service provided by ICRCL would be diminished if the Committee failed to take proper account of new information on the way in which contaminated land is best developed and it is to be expected, therefore, that advice offered by the Committee undergoes some modification as time goes by. However, the extent to which real changes in approach, as opposed to subtle differences in interpretation and terminology, have taken place is open to question although in some recent publications the Committee has attempted to suggest that the current approach advocated by ICRCL has evolved in response to what it regards as unsatisfactory or misguided redevelopment attempts in the past.

An example is provided by the proposal of a new operational definition of contaminated land which has already been mentioned briefly (see page 2). According to ICRCL, the older commonly accepted definition suffers a number of flaws which cause it to be 'very general and unhelpful in practice' (5). These include that it does not fully explain the source or nature of the contamination, that it infers many contaminated sites are unusable when there is little or no risk for some types of development and most importantly, that the definition fails to stress the need to assess the hazard posed against the proposed use of the site. The Committee states that these inadequacies have led to 'over-reaction' in the past and a lack of attention as to how the possible risks can be assessed and hence avoided or overcome. In an effort to correct these deficiencies the new definition is drawn strictly within the context of redevelopment and contains a strong reference to the need for a critical assessment of the site in relation to the proposed development. The definition also refers to the fact that remedial action can include changing the form or the layout of the development.

A second area of interest is the provision of guideline information on levels of contaminants in soils. The important principle here is to decide on those levels of contaminants on a site which represent no significant hazard to the users, occupiers or other target groups in contact with the site, and those which, being of a sufficient quantity or concentration, require some action to be taken in connection with the site. This type of information is of obvious value to Local Authorities who are required to make decisions regarding the development of contaminated land, and during 1979–1983, ICRCL has issued a number of guidance documents. While the advice provided has always been of a tentative nature the terminology, content and presentation has changed, particularly in recent times. Early ICRCL guidance used the argument of 'acceptable' levels of contaminants in which three different scenarios were envisaged. If, after a thorough site survey, observed levels of contaminants were found to be below the guideline or acceptable levels of contaminants then it could be assumed that the presence of the contaminants posed no significant hazard to critical groups and no subsequent action would be required. If observed levels slightly exceeded guideline levels then the situation could be considered to be 'undesirable'. In this case, it would be appropriate to consider, taking into account all available information, to what extent, if any, remedial action should be taken with regard to the proposed development. At some still higher level of contamination, the situation could be described as being 'unacceptable' for a particular use and remedial measures appropriate to that use would then become essential. Early guidance did not attempt to define any level of contamination other than the 'acceptable' level but suggested that there was a need to use professional judgement and expertise in those situations where guideline levels were exceeded.

Current advice uses a different terminology and a slightly different format but it retains the essential features of earlier advice. ICRCL 59/83 was published in May 1983 and formally replaces all former guidance of this type. Guideline information is presented in the form of 'trigger concentrations' which, together with an appreciation of local natural background levels, are intended to be of assistance during the interpretation of data derived from site investigation. As before, three different situations are discussed. In those cases where observed levels of contaminants are below the trigger concentration value, the site can be regarded as being uncontaminated in so far as the presence of the contaminants poses no additional risk to the critical group. On those occasions where trigger concentrations are exceeded, ICRCL 59/83 states:

> If the trigger concentrations are not greatly exceeded, remedial action is not automatically required. At some still higher concentrations the risks posed by the presence of contaminants might be judged unacceptable for a particular use. Remedial measures appropriate to that use would then become essential or another use, less 'sensitive' to the contamination would have to be adopted.'

The philosophy behind the use of trigger concentration values and the role of professional judgement and expertise have been described as follows:

> If '*x*' is the trigger concentration for a given contaminant then although '$x - 1$' can be regarded as safe '$x + 1$' is not necessarily dangerous, unacceptable or undesirable. The trigger concentration only represents a value at which consideration needs to be given to the presence of the contaminant: professional judgement and experience than decide whether or not remedial action is necessary (7).

It would appear that ICRCL now considers it necessary to stress that in those situations where guideline levels are exceeded there should not be an automatic assumption that a site is dangerous and therefore unusable, and moreover that remedial measures are essential only if the proposed development is incapable of modification or if alternatives do not exist. It will be appreciated that current guidance, in respect of the use of guideline information, is wholly consistent with the aims of the new definition of contaminated land referred to above.

Some differences exist between this and earlier forms of guideline advice including the simplification of after-use categories and attendant guideline levels of contaminants, the withdrawal of advice on some inorganic constituents such as barium, fluoride and antimony and the inclusion of guidance on some organic substances. However, the principles which govern the use of this type of information have remained largely unchanged and these include:

(a) Guideline information applies only in those cases where a thorough site investigation has been carried out
(b) Guideline levels should not be used to indicate maximum permissible levels
(c) Guideline levels should relate to sampling and analytical procedures. In the case of ICRCL 59/83 these are specified as concentrations measured on 'spot' samples rather than averaged, bulked or composited samples. The use of maximum rather than mean, model or median values is recommended
(d) Guideline information should aim to accommodate the exercise of professional judgement.

An important point to note is that although current ICRCL advice is based on fewer land-use categories than was previously the case, when different classes of domestic garden based on size were specified, the concept that advice should take into account the different requirements of particular types of redevelopment has been retained. It has always been recognized that some uses of a site are more sensitive to the presence of contamination than others. In general, a sensitive use of a site would be one which included the provision of housing with gardens in which there may be prolonged and close contact with the soil, as may be the case with small children, or where there is a possibility of regular consumption of home-grown produce (see Chapter 3). The least sensitive uses of a contaminated site would include those developments in which large areas of the soil are covered by buildings, concrete or tarmac such that access to the underlying contaminated soil is restricted. While there may be short-term risks involved in these types of developments, for example during building operations, and/or a requirement to protect buildings and services over the long-term, there is nevertheless a much reduced risk of direct contact between contaminants and the subsequent users of the site.

It will be clear from the preceding sections that current ICRCL policy is concerned with promoting the redevelopment of contaminated land for beneficial use. Local Authorities and others with a responsibility for contaminated land are urged to adopt a systematic approach involving the close scrutiny of a contaminated site in relation to the proposed development. The careful assessment of information derived from a properly conducted site survey, and other sources, is then used to decide whether a redevelopment proposal goes forward unmodified, whether remedial measures are required or whether an alternative use of the site would be more appropriate. ICRCL recommends 'the form of the development should preferably only be decided after assessing the site, and should be chosen so as to minimize the hazards to the users' (5). This type of approach seeks to reduce reliance on the use of physical, chemical or biological means of ameliorating conditions on a site and could reasonably be described as being one of 'making the development fit the site' rather than one which is concerned, using remedial measures where necessary, with 'making the site fit for the development'. In many cases, development has to proceed under less favourable circumstances to those recommended by ICRCL and for this reason a large section of this book is devoted to considerations of remedial or reclamation methods.

There are relatively few specific examples of what the response to current Government policy in this area has been. There are, however, indications that difficulties are being experienced with the interpretation of trigger concentration data particularly in connection with their relationship with local background levels of contaminants. Trigger concentrations are intended to indicate the levels of contaminants below which a site may be regarded as uncontaminated and they therefore represent a level of contamination at some point above typical background concentrations. However, there are examples in which typical levels of metallic contaminants, over large areas of land, exceed the trigger concentrations (see page 5). The question must then be raised as to whether trigger concentrations have been set too low or whether guideline information must be interpreted more flexibly in those areas which, for one reason or another, have inherently high levels of contaminants in the soil. There may be some merit in developing 'reference levels' of contamination for urban environments by which it is possible to judge whether conditions on one site are better or worse than in other similar areas (41). However, in the absence of a national set of reference levels ICRCL advice to take into account local natural background levels of contaminants during the interpretation of site investigation data has met with some criticism. According to one report of a recent Institute of Environmental Health Officers (IEHO) meeting to discuss the contents of ICRCL 59/83 (42), this aspect of policy was rejected on the grounds that it may lead to different standards of care in different parts of the country. It was also reported that EHO's are concerned that even after several years of debate, guideline information is still of a tentative nature leaving much of the responsibility for making decisions on contaminated land with Local Government departments. The report went on to suggest that current DoE policy on development matters in general, which, it was stated, puts pressure on Local Authorities to release land for development, may have influenced ICRCL to urge Local Authorities not to over-react to contaminated land. This account of the meeting was subsequently challenged (43) when it was pointed out that despite their tentative nature, the trigger concentration values provided a base from which to work and furthermore, that ICRCL advice was to be extended to include categories of 'unacceptable' and 'undesirable' levels of contaminants. Some progress in this area has been reported (44) in relation to lead and cadmium and an extension of guideline information in this direction will do much to allay the fears of EHO's. However, there has been no response to the suggestion that Departmental policy on development matters in general has influenced policy on contaminated land in particular. It is certainly the case that the system used until recently for the allocation of Derelict Land Grant, under the so-called Category A scheme, puts pressure on Local Authorities to submit reclamation schemes which will not only attract substantial private investment but which also propose the use of land for a small number of qualifying uses including housing. Any suggestion that difficulties may arise if contaminated land is reclaimed for such purposes would have two negative effects—firstly it would make such schemes less attractive to private developers, as appears to have been the case in South Yorkshire where technical problems associated with past mining activities have deterred private investment (45), and secondly it would deprive Local Authorities of much needed funds to carry out reclamation work. The whole question of finance for the redevelopment of contaminated land is an important one and the issue is explored in greater detail in Chapter 9.

In order to conclude this section on the Central Government response to contaminated land it is necessary to refer to recent reports which suggest that Governmental stringency in relation to Departmental resources has affected the ability of ICRCL to fully implement its original programme of work. There has been a recent analysis of the contents of the first four Management Information Systems for Ministers (MINIS) Reports which were introduced into the DoE in 1979 as an aid to improve the efficiency of management practices within the Department (46). According to this analysis, the Central Directorate on Environmental Protection (CDEP), within which ICRCL operates, has suffered a 26% cut in staff over the period 1980–1984—a figure roughly double that of the average level of staff reduction in the Department as a whole. The report states that two important functions of CDEP, as they relate to contaminated land, have been affected. These are a long delay in the publication of the Pollution Paper on contaminated land (originally set for 1980, and now apparently abandoned altogether) and the reduction of the advisory service to Local Authorities. A further analysis (47) of the contents of the fifth MINIS Report, published in August 1984, commented that while some staff reductions in the Department have been reversed, the position with regard to CDEP has remained unchanged. These findings have been endorsed in the tenth report of the Royal Commission on Environmental Pollution (48) which expressed fears that the ability of CDEP to act as an effective central source of advice on UK environmental policy had been adversely affected by a reduction in the allocation of resources.

It is interesting to note that the Royal Commission has recently completed a study concerned with pollution by wastes, including the problem of contaminated land. The Commission originally invited comments on the following areas in relation to contaminated land:

(a) Avoidance of future land contamination problems from existing industrial activity and on site waste disposal and storage
(b) Problems of decommissioning and of business failures
(c) Technical problems relating to the legacy of existing contamination
(d) Avoidance of future land contamination problems from mining activities

Both written and oral evidence were considered by the Commission and a report has recently been published (49). It was decided that contaminated land did not merit extensive consideration on the grounds that it largely results from past and generally obsolete practices. While the RCEP decision will be regarded with some disappointment, not least by those Local Authorities acknowledged by the Commission to be directly affected by contaminated land, two important recommendations were made. Both are intended to facilitate the early identification of this category of land. The first is a suggestion that the DoE should issue a circular outlining the circumstances in which land contamination would be a material consideration in refusing, or otherwise defining, planning permission and encouraging liaison between planning and environmental health departments on relevant planning applications. The second is concerned with the role which could be played by the land registration system in making information concerning land contamination more widely accessible.

1.2.2 *Local government response*

Local Authorities have made an important contribution to the body of knowledge on contaminated land. The statutory responsibilities of Local Authorities in areas such as planning and development, health and safety, pollution control and waste disposal have meant that many Local Authorities have had an unavoidable involvement in contaminated land in one capacity or another and much of the practical responsibility for managing contaminated land has fallen to Local Governmment departments.

Many Local Authorities, either on a departmental basis or through the offices of specialist teams, have developed procedures and policies for dealing with contaminated land and have presented accounts of their experiences (50–52). The approach developed by the Greater London Council (GLC) has been particularly well documented over the years and will be reviewed briefly since the GLC experience provides a good example of how many Local Authorities have responded to the challenge of contaminated land.

In common with other Local Authorities the GLC first became involved with contaminated land in the early 1970s when former industrial land was acquired for development purposes. An early account of the approach taken by the GLC to such land was provided in 1976 by Chapman, Baker and Burns (53). The authors defined the hazards associated with the land and the groups thought to be at risk. In addition the procedures used by the Authority at that time were outlined and the first set of GLC tentative guideline values were presented. A second paper, published by Kelly in 1979 (54) gave an up-dated account of GLC practice in respect of contaminated land which indicated where the Authority had modified its approach as experience was gained. Further hazards and critical groups had been identified and sampling procedures modified. For example dug pits rather than boreholes had become the preferred method of sampling contaminated ground and analytical procedures had become more selective. Kelly described three aspects of the GLC approach in detail. These included the use of guideline data on contaminant levels, the scoring system for wastes arising from the development of contaminated land and the organizational procedures used by the Council including a Code of Practice to enable officers to deal more effectively with contaminated sites. More recent accounts of the GLC approach were presented to a one-day symposium entitled, 'Contaminated Land – the London Experience' (55) in which papers covering current procedural arrangements, site investigation and assessment of results, the problems of methane and the disposal of wastes arising from the development of contaminated land were given. The proceedings of this conference have been published in the form of a supplement to the recently launched GLC publication, *The London Environmental Bulletin* which also carries regular features on GLC activity in respect of contaminated land such as the latest GLC guidance on site investigation (56) and methane (57). This relatively recent initiative must be viewed against an established tradition of the GLC to publish practical accounts of its experience in relation to contaminated land. There is already an extensive literature which descibes GLC involvement in the assessment of various types of contaminated land including former Ministry of Defence land, waste disposal sites, former dockland and redundant gasworks land (see for example 13, 37).

Clearly, the GLC has acquired considerable technical experience of managing contaminated land and has played a valuable role in developing procedures and policies of practical application as seen by the GLC representation on BSI Technical Committee EPC/47. Moreover by demonstrating a strong commitment to publish accounts of its experiences the GLC, along with other Local Authorities, has made a valuable contribution to the present level of understanding as to what constitutes contaminated land and how the technical problems associated with redevelopment can be approached.

1.3 National response to contaminated land in Europe and North America

1.3.1 *Introduction*

Since contaminated land results from the use of land for industrial or waste disposal purposes it is only to be expected that such land will be a feature of all industrialized nations, and may also exist in embryonic form in some developing countries. However the perception of contaminated land and the hazards it presents is likely to vary between nations according to various social factors including domestic policies on land use development and the age, structure and location of industrial activities. Owing to its rather special constitution the NATO/CCMS pilot study on contaminated land has afforded a useful insight into how a number of countries regard the problem of contaminated land. It is apparent that views differ not only as a result of the social factors outlined above but also in response to some of the more sensational pollution episodes which have been experienced by participating countries.

1.3.2 *The Netherlands*

In the Netherlands land is traditionally regarded as a scarce and precious resource. There is a general recognition of the need to protect the multifunctional properties of soil and the quality of groundwater upon which drinking water supplies are frequently dependent. The presence of contamination in soils is therefore of particular concern and it is apparent that the Dutch Authorities are currently making vigorous efforts to locate and treat contaminated land. While there are indications that there are a number of contaminated sites of some concern in the Netherlands, one particular major incident, the housing development at Lekkerkerk, may have had considerable influence on the approach taken by the Authorities to the problem of contaminated land.

A detailed account of the circumstances under which the Lekkerkerk incident arose, and the remedies applied, is available (58) and only a brief summary will be given here. During the 1960s the Lekkerkerk district authorities were faced by two pressing community problems; a lack of cheap residential development land to house an expanding post-war generation and a need to rationalize waste disposal practices as the volume of materials generated by this population grew. An attempt was made to resolve the problem by reclaiming land using both domestic and industrial waste to fill the network of ditches which had formed the drainage system of the land. Although the potential engineering problems of building houses on such land were anticipated the implications with respect to possible contamination were not fully realized until after the housebuilding programme was completed in 1975. During the next three years, abnormally high numbers of burst pipes were recorded and in 1978, a workman repairing a gas main became unwell when polluted water flooded a working shaft. Subsequent examination of the area revealed the presence of xylene and toluene in the creep spaces beneath houses (benzene was also detected at one stage), the contamination of drinking water supplies and mechanical weakness in the pipes serving the houses. The most serious effects were associated with houses built directly over the filled ditches and test excavation of the fill materials provided samples of highly polluted soil and indicated the presence of broken and partially full containers. The findings of the investigation confirmed fears that chemical wastes from the dyestuffs industry had been deposited. High levels of pigments and lower levels of cleaning agents were also found. The Local Authority immediately introduced short-term corrective measures including the provision of bottled drinking water and advice to residents to ventilate the creep space beneath their homes. However, following Central Government involvement, residents were moved out of their homes to allow the excavation of contaminated soil and the replacement of uncontaminated material.

The entire clean-up operation, from confirmation of soil contamination in early 1980 to completion by the end of 1981, involved the removal of 153 000 t of soil and the demolition and rebuilding of a school and gymnasium . While a commendable achievement in terms of technical and organizational skill, the operation was also extremely expensive, incurring costs in the region of $ 65 million (US) (59). However, perhaps one of the most important effects of the Lekkerkerk experience was the introduction of a broad Government strategy aimed at remedying existing contaminated sites and preventing the occurrence of similar sites in the future. There are three elements to this Government strategy:

(a) A national survey of contaminated sites
(b) The introduction of legislation and the provision of finance
(c) The publication of technical information.

In order to determine the extent of contaminated land in the Netherlands a national survey of contaminated land was carried out in 1980. The survey indicated the existence of over 4000 sites including 1000 in which some form of corrective action is required (59).

Pending the introduction of a comprehensive Soil Protection Act (see below) the Interim Soil Reconstruction Act (also referred to as the Soil Clean-up (Interim) Act) was passed in 1983 under which it became possible to identify and treat contaminated land. Under the Act, Provincial Authorities are required, on an annual basis, to submit soil reconstruction plans covering a five-year period, to the Minister for Housing, Physical Planning and the Environment. In respect of the legislation both Central and Local Government have far-reaching powers. For example, the Minister can demand user or ownership rights to a site in order that remedial work can be carried out. There are also provisions to oblige owners or occupiers of a contaminated site to co-operate in soil treatment operations, and powers to shut down an offending process or allow it to continue only under specified conditions. Both Provincial and Local Government are authorized to establish temporary storage sites for contaminated material arising from remedial work. The legislation has attempted to embody the principle of 'the polluter pays' and there are formulae to establish the costs to industry where appropriate. However, in recognition of the difficulties of proving liability, State funds are available in the first instance to support reclamation work. There are also clauses in the Act to deter speculation in contaminated land where it may be purchased at low cost, and after a State clean-up, resold at high cost. The Soil Protection Act, which was due to come into operation in 1985, contains sections to replace the provisions of the Interim Act. The Soil Protection Act is designed not only to protect soil resources from pollution in the future but it is also expected to allow for the treatment of less severely contaminated sites which are not covered under the Interim Act.

A variety of programmes are available in the Netherlands in which State funds can be used to finance clean-up operations in addition to research into appropriate techniques (60). Central Government, together with Local Government funds which are available for the enactment of the Interim Soil Reconstruction Act amount to an annual expenditure, up to and including 1987, of approximately 165 million guilders (equivalent to about £39 million). For the fundamental development of remedial techniques, 1 million guilders is available annually until 1988 from the Department of Housing, Physical Planning and the Environment while under the Environmentally Beneficial Technology Stimulation Scheme, approximately 7 million guilders is potentially available until 1986. Under a related programme, known as the Biotechnology Innovation Research Programme, set up under the Ministry of Economic Affairs, nearly 3 million guilders per annum have been set aside for environmental applications.

Technical information has also been prepared by Central Government to aid the work of soil remediation. One example is the *Handbook of Soil Reconstruction Techniques* which provides a readily accessible source of information on the technical, financial and environmental aspects of various remedial techniques (61). A second manual deals with safety aspects of the investigation of contaminated sites and covers such items as the statutory obligations of the body commissioning site investigation work, general organizational procedures and details of safety provision which must be made during the investigation stage.

Current Dutch policy on the treatment of contaminated land favours those methods which remove the contamination from the soil. Containment or excavation and disposal options would only be pursued if the former course of action were to be ruled out on the grounds of prohibitive cost or unacceptable danger to public health. Accordingly much research effort has been concentrated on methods of on-site processing of contaminated soils including those involving thermal treatment of soils to destroy organic contaminants, extraction procedures using aqueous and organic solvents and microbiological treatments which remove contaminants by biodegradation. Recent accounts of the effectiveness of on-site methods (59, 62, 63) conclude that while much work remains to be done to establish the effectiveness of the treatments on a large scale, they do have valuable potential application and are deserving of further investigation.

1.3.3. *Federal Republic of Germany*

In common with other European countries, West Germany has also had to deal with the consequences of the industrial use of land (64). Examples of contaminated land include a former gasworks site (cyanide and phenolic residues), a chemical waste dump (hexachlorocyclohexane (HCH)), an abandoned chemical waste disposal site (unknown chemical wastes contained in corroded and leaking drums) and a former factory site (cyanide contamination). Land contaminated as a result of metal smelting and land to which contaminated sewage sludge and river sediments have been applied for agricultural purposes have also presented problems for the German Authorities. In addition, there are examples of severe groundwater contamination by arsenic, tetrachloroethylene, fuel oil and other organic derivatives resulting from the use of land for industrial purposes including the storage of materials (65).

A variety of remedial techniques have been used to ameliorate the conditions on these sites including the excavation of contaminated soil for disposal to chemical waste landfill, incineration, encapsulation, deep ploughing (where this is not considered hazardous to groundwater sources) and simple liming of soil to reduce metal mobility. In those cases of groundwater contamination both chemical and biological techniques have been used to treat soils and leachates. Encapsulation and treatment of leachates are the proposed methods of managing a hazardous waste landfill site in which approximately 370 t of dioxin containing waste have been deposited (65). In this case, despite there being no evidence of dioxin in groundwater, soil, plant or animal samples, public reaction to the presence of dioxin at the site has influenced the decisions taken with regard to the site.

The Laender Authorities carry the main responsibility for contaminated land (66) and although there has been recent agreement between the Federal and Laender Authorities on a common strategy, to date each Laender has proceeded on an individual basis with some developing special programmes to deal with contaminated land. For example, in North Rhine Westfalia special arrangements exist for the identification and treatment of redundant hazardous waste sites (Altlasten or historical burdens). While there is a common path of registration, investigation, monitoring and, if necessary, remedial action there is no uniform evaluation system between Laender Authorities for judging the extent of the hazard posed by a particular site. For example, while Hamburg uses a system roughly corresponding to the United States Environmental Protection Agency's Hazard Ranking System, other Authorities use local standards to judge the extent of any environmental threat associated with a site. Despite these differences in approach national estimates of the number of problematic hazardous waste disposal sites, and contaminated industrial sites have recently been given as 3000 and 2500 respectively.

A number of working groups are reviewing the problem of contaminated land in the Federal Republic of Germany, most notably the Laender Working Group on Wastes (Landerarbeitsgemeinschaft Abfall, LAGA) which is currently developing appropriate national guidance on contaminated land which will include formal criteria on monitoring and sampling. Other groups are active in the field of groundwater protection and engineering aspects of remedial treatment.

Financial support for remedial work comes from a number of sources. In North Rhine Westfalia for example the treatment of contaminated land is supported under existing Authority budgets while in other areas, special funds are available to help in the reclamation of derelict, and frequently contaminated, industrial land. In other cases companies which have been associated with the contamination present on a site have met the cost of cleaning-up (see for example the HCH contaminated site in which a chemical company spent nearly 200 000 DM for remedial works (65)). However, as has been the case in the Netherlands, the application of the 'polluter pays' concept is seldom a straightforward matter and the German Authorities have examined the possibility of setting up a fund, along the lines of the US 'Superfund' to finance treatment works.

Research into remedial measures for use on contaminated sites has been made a priority item by the Federal Minister for Research and Development. Areas of work for inclusion in the Environmental Research and Development Programme (1984–1987) are:

Development and testing of investigation and evaluation procedures
The use of remedial measures including in-situ and on-site methods of treating contaminated soils
The protection and treatment of groundwater—degasification and treatment of landfills
Long term performance of remedial measures.

In all approximately 80 million DM are available for research and development in these areas.

1.3.4 *The United States of America*

The use of former industrial and waste disposal land for development purposes, including housing, is not unknown in the USA. A detailed account of three such developments has recently been given (67) in which a former landfill site, a redundant steel works and land previously occupied by an explosives manufacturer have been used for residential purposes, a hotel/office/commercial complex and a 'new-town' development respectively. These examples were used to demonstrate that contaminated land can be redeveloped successfully provided adequate precautions are taken and in all cases detailed site investigation, and frequently extensive remedial measures, were carried out before development proceeded. However concern has been expressed that some types of development, for example on former gasworks sites (63) have not proceeded in a proper manner, or have taken place before the environmental significance of redeveloping contaminated land were fully realized (69).

In the examples described, development has come about as a result of high demand for land in established urban areas or on the basis of a favourable economic return on investment. In general, however, land resources in the United States are such that, compared with the more densely populated countries of Europe, there is little pressure to use old industrial sites for development purposes. As a result, American policy on contaminated land in the past was geared more towards the control and containment of potential hazards, particularly with respect to the many 'uncontrolled hazardous waste sites' (3) rather than one dominated by considerations of the implications for redevelopment. However, there is now a growing emphasis placed on remedial methods which eliminate the problems associated with uncontrolled hazardous waste sites with control and containment viewed increasingly as only the first stage of a total remedial treatment.

One incident which did involve the redevelopment of land adjacent to a hazardous waste site and which may have exerted considerable influence on the national perspective of contaminated land, involved the Love Canal toxic waste site at Niagara Falls. The site achieved some notoriety in the late 1970s after the clay capping used to contain the wastes was breached during housebuilding operations. Subsequent ingress of water into the site caused flooding of house basements with contaminated water. Concern over the health implications for residents was so intense that at one stage families were evacuated from the area. The Love Canal site became the subject of a number of highly controversial reports which presented conflicting assessments of the hazards posed by the contaminants (70, 71) and it appears that the main lesson to have been learnt from the episode is that extreme caution should be exercised when developing sites of this nature for residential use in particular.

The official response to sites of this kind was to introduce legislation designed to protect the land environment in the future. The Resource, Conservation and Recovery Act (RCRA) was introduced in 1976 to provide a means of ensuring the safe handling of waste materials at all times. However RCRA did not provide Federal and State Government with the authority to respond directly to problems caused by existing hazardous waste sites. The Comprehensive Environmental Response, Compensation and Liability Act (CERCLA), commonly referred to as the 'Superfund' Public Law 96–510, was passed in 1980 to provide this authority, not only in times of emergency, but significantly in those cases where a long-term solution to problems is required. CERCLA also provided a means of financing investigation and remedial work in connection with sites which pose a threat to public health or the environment. Financial support is provided by means of a Trust Fund (Superfund) which derives 86% of its income from taxes on the petroleum industry and forty-two other feedstock chemical industries and the remainder from Federal sources. Over the five-year lifetime (1980–1985) of the Superfund programme approximately $1.6 billion is available, and there is evidence that the budget may be increased by a factor of six, although conditions are attached to such an increase as explained below.

A full account of the operation of the Superfund law is available (72) and only a brief description will be provided here. Action taken under CERCLA results from co-operation between the United States Environmental Protection Agency (EPA) and State Authorities and the programme operates according to a National Contingency Plan. This allows the Authorities to take action in respect of a site in three main ways:

(a) Immediate removal—emergency action in response to an immediate hazard to public health or the environment
(b) Planned removal—an intermediate response when a serious but not immediate hazard is involved

(c) Remedial response—intended to provide a long-tern solution to problems arising from a hazardous waste site.

CERCLA legislation is used to some advantage in the identification of hazardous waste sites since it requires owners and operators to notify the EPA of their activities. Sites already known to EPA and submissions from the public also assist in the location of sites. Sites are then evaluated using all available data and field investigation. Finally the sites are ranked according to a hazardous ranking system employing various criteria such as risks to human health, to drinking water supplies and to sensitive ecosystems. CERCLA requires the compilation of a National Priorities List (NPL) to include sites which qualify for the remedial response option. The NPL published by EPA in October 1984 (73) listed 538 sites which meet the criteria for action under Superfund and a further 248 sites proposed for inclusion on the NPL. Information is given on the scores allocated to each site according to the hazardous ranking system, those sites designated by individual States or Territories as being the top priority site and details of the status of sites in relation to the type of response taken and progress made.

Once a priority site is identified, various courses of action can be taken as follows:

(a) The party responsible for the site can treat the site
(b) The State or Local Government body can undertake all or part of the responsibility of dealing with the site without recourse to Superfund monies
(c) Legal action can be taken to force the responsible body to treat the site
(d) Financial resources from the Trust Fund can be used to treat the site but only in those cases where alternatives (a)–(c) cannot be implemented.

EPA may then proceed under the Superfund legislation and, at a later date, seek to recover costs by legal action. While this course of action may help to prevent delays in bringing a site under control, it is clear that the legal procedures involved are extremely complex and may themselves demand substantial financial resources.

When a remedial response option is selected the aim is to decide on 'the least expensive remedy that is technologically feasible and reliable and that effectively reduces the danger and adequately protects public health, welfare and the environment'. Five basic procedures are used in the planning of a remedial response the first of which is the remedial action master plan. This comprises all the information available on which to decide the possible types of remedial response which fulfil the objectives described above. At this stage an opportunity is also provided to the State to decide on its role in any subsequent action. CERCLA provides for two arrangements:

(a) A co-operative agreement in which the State takes the lead role and is responsible for planning and executing the remedial programme although the EPA maintains a monitoring role

(b) Superfund State Contract in which the EPA takes the lead role and the work is carried out using EPA Superfund contractors or the Army Corps of Engineers.

Under CERCLA the State must share the costs of remedial action (at least 10% if the site was privately owned, and 50% if publicly owned at the time of hazardous waste disposal). States must pay for operation and maintenance costs (except the first six months when EPA finances 90% of these costs), and they must provide approved facilities for the treatment of any materials arising from the site during treatment.

The next three stages of planning the remedial response involve the remedial investigation, feasibility study and selection of the remedy during which time a preferred course of action is identified. Finally a remedial design, usually in the form of a set of design plans, is submitted for tender.

Clearly, the Superfund legislation represents a challenging and ambitious means of reducing the hazards arising from a major source of contaminated land. However, as may be expected for a relatively new addition to Federal environmental policy, the Superfund programme has been criticized on a number of grounds. It has been claimed for example that progress made under the legislation has been too slow. Approximately 18 000 potentially hazardous sites were known to EPA in 1984 and yet it is reported that only six sites had been cleaned-up in the four-year operation period of the Superfund programme (74). The proposed increase to the EPA budget mentioned previously has conditions attached which are intended to rectify the situation by calling for remedial action start-ups at a rate of 150 sites per year as well as more rapid progress on EPA study effort. The allegation that only six sites have been cleaned up is refuted by EPA (75) who claim that under both the emergency and remedial response programmes, work on between 500 to 600 sites had begun with more than 150 sites completed and a further 150 stabilized. The six sites in question refer to those which have been removed from the National Priorities List indicating the completion of the remedial works including monitoring. The EPA point out that some sites may remain on the NPL for five to ten years while others may remain on the list indefinitely as they will require long-term monitoring to ensure that remedial measures remain effective. On the question of increased financial resources the EPA maintain that the increased costs which would be entailed by complying with the suggested rate of start-ups would have the effect of hampering, rather helping clean-up operations.

Administrative aspects of the Superfund programme have also been the subject of criticism (76). It is claimed, amongst other things, that too much bureaucracy is involved, that there is a lack of delegation within the EPA between central and local levels, and that some delays in the programme are attributable to the use of only a small number of contract laboratories for analytical work.

Finally, it appears that there is some evidence of basic inconsistencies between CERCLA, and the RCRA legislation of 1976 which was intended to prevent current hazardous waste sites from becoming the 'Superfund' sites of the future (77). The main defects of the RCRA legislation are stated to include inadequacies in the sampling and monitoring procedures used on both permitted and interim status hazardous waste sites, the lack of provision to ensure that leaks are properly attended and repaired and no requirement for financial assurance from the operator or owner of a site to cover the costs of corrective action. There is, however, a means of dealing with this last defect in that CERCLA does provide for a post-liability trust fund which comes into effect thirty years after a site operator has closed and monitored his site. At this point, the site can be handed over to the Federal Government for perpetuity. If the site fails after the Federal Government has assumed responsibility for it the remediation will be paid for out of the post-liability trust fund.

References

1. Page, R.A. (1984) Overview of hazardous waste site problems in Wales. In: *Proc. Conf., Management of Uncontrolled Hazardous Waste Sites, Washington DC*, HMCRI, Silver Spring, Maryland, 594–597.
2. Interdepartmental Committee for the Redevelopment of Contaminated Land (ICRCL) (1983) Guidance on the assessment and redevelopment of contaminated land. ICRCL Paper 59/83, 1st edition, DoE, CDEP, London.
3. Smith, M.A. (1985) In: *Contaminated Land*, Report of the NATO/CCMS pilot study on contaminated land, Plenum, New York, 1–11.
4. Anon. (1983) Draft British Standard Code of Practice for the Identification and Investigation of Contaminated Land. Ref. 83/55992, British Standards Institution, London.
5. Beckett, M.J. and Simms, D.L. (1984) The development of contaminated land. In: *Proc. Conf., Hazardous Waste Disposal and the Re-Use of Contaminated Land, London*, SCI, London.
6. Eustace, J. (1984) Analysis and assessment of contaminated land. *London Environmental Supplement*, No. 7, 7–9.
7. Beckett, M.J. (1983) A systematic approach to the redevelopment of contaminated land. In: *Proc. Conf., Reclamation '83, Grays*, Industrial Seminars Ltd., Tunbridge Wells, 555–563.
8. Barry, D.L. (1982) Prospective surveys for contaminated land. In: *Proc. PTRC Annual Summer Meeting*, University of Warwick, Coventry.
9. JURUE (1981) Contaminated sites in the West Midlands; a prospective survey. (Unpublished report to the Department of the Environment.)
10. Dumbleton, M.J. (1980) Historical investigation of site use, In: *Proc. Conf., Reclamation of Contaminated Land, Eastbourne*, 1979. SCI, London, 1980. B3/1-B3/13.
11. Dumbleton, M.J. (1983) Air photography for investigating natural changes, past uses and present condition of engineering sites, TRRL Report 1085 (Department of the Environment), Department of Transport, TRRL, Crowthorne.
12. Joyce, F.E., Williams, H.E. and Haines, R.C. (1982) Land contamination and urban development. In: *Proc. PTRC Annual Summer Meeting*, University of Warwick, Coventry.
13. Lowe, G.W. (1984) Investigation of land at Thamesmead and assessment of remedial measures to bring contaminated sites into beneficial use. In: *Proc. Conf., Management of Uncontrolled Hazardous Waste Sites*, Washington DC., (HMCRI, Silver Spring, Maryland) 560–564.
14. Heeps, K.D. (1982) The reclamation of a disused sewage works, *Public Health Engr.* **10**, 213–218.
15. ICRCL. (1979) Progress Report of the Interdepartmental Committee on the Redevelopment of Contaminated Land, ICRCL Paper 19/79, DoE, CDEP, London.

16. ICRCL. (1980) Notes on the redevelopment of scrapyards and similar sites. ICRCL Paper 42/80, 2nd edition, DoE, CDEP, London.
17. ICRCL. (1983) Notes on the redevelopment of gasworks sites. ICRCL Paper 18/79, 4th edition, DoE, CDEP, London.
18. ICRCL. (1983) Notes on the redevelopment of sewage works and farms. ICRCL Paper 23/79, 2nd edition, DoE, CDEP, London.
19. ICRCL. (1983) Notes on the redevelopment of landfill sites. ICRCL Paper 17/78, 5th edition, DoE, CDEP, London.
20. ICRCL. (1984) Notes on the fire hazards of contaminated land. ICRCL Paper 61/84, 1st edition, DoE, CDEP, London.
21. ICRCL. (1985) Asbestos on contaminated sites. ICRCL Paper 64/85, 1st edition, DoE, CDEP, London.
22. Lepp, N.W. and Harris, M.R. (1980) A strategy for evaluation of soil covering techniques to reduce trace metal uptake by soft fruits and vegetables. In: *Proc. Conf., Reclamation of Contaminated Land*, Eastbourne, SCI, London. C7/1—C7/9.
23. Jones, A.K., Johnson, M.S. and Bradshaw, A.D. (1980) Biological aspects of the treatment of heavy metal contaminated land for housing development schemes. *Ibid*, C3/1–C3/9.
24. Jones, A.K., Johnson, M.S. and Bell, R.M. (1981) The movement of metals in reclaimed metal contaminated land. In: *Proc. Conf., Heavy Metals in The Environment*, Amsterdam, CEP Consultants, Edinburgh, 375–382.
25. Jones, A.K., Bell, R.M., Barker, L.J. and Bradshaw, A.D. (1982) Coverings for metal contaminated land. In: *Proc. Conf., Management of Uncontrolled Hazardous Waste Sites*, Washington DC., HMCRI, Silver Spring, Maryland, 183–186.
26. Harris, M.R., Harrison, S.J., Wilson, N.J. and Lepp, N.W. (1983) Elemental composition of fruit crops and their use as bioindicators of reclamation treatment performance. In: *Proc. Conf., Heavy Metals in the Environment*, Heidelberg, CEP Consultants, Edinburgh, 1162–1165.
27. Harris, M.R., Harrison, S.J. and Lepp, N.W. (1984) Seasonal variations in the metal content of grass and its use as an indicator of reclamation treatment performance. *The Science of the Total Environment*, **34**, 267–278.
28. Cairney, T. (1982) In-situ reclamation of contaminated land: The problem of a safe design. *Public Health Eng*. **10**, 215–218.
29. Cairney, T. (1984) Reclaiming chemically contaminated land for housing and other sensitive uses. In: *Proc. Conf., Environmental Contamination*, London, CEP Consultants, Edinburgh, 294–299.
30. Bell, R.M. and Parry, G.D.R. (1984) The upward migration of contaminants through covering systems. In: *Proc. Conf., Management of Uncontrolled Hazardous Waste Sites*. Washington DC., HMCRI, Silver Spring, Maryland, 588–591.
31. Cairney, T. (1986) Soil cover reclamation experience in Britain. In: *Proc. Conf., Contaminated Soil*, Utrecht, Netherlands, (Martinus Nijhoff, Dordrecht), 601–614.
32. Environmental Advisory Unit, University of Liverpool (1983) Movement of metals in reclaimed metal—contaminated ground-site survey report. (Unpublished report to the Department of the Environment.)
33. Beever, P.F. (1986) Assessment of fire hazards in contaminated land. In: *Proc. Conf., Contaminated Soil*, Utrecht, Netherlands, 1985, (Martinus Nijhoff Publishers, Dordrecht), 515–522.
34. Wilson, D.C. and Stevens, C. (1981) Problems arising from the redevelopment of gas works and similar sites, AERE Report R10366, HMSO, London.
35. Barry, D.L. (1982) Treatment options for contaminated land. Report for the Department of the Environment, Atkins Research and Development, Epsom.
36. Anon. (1980) *Proc. Conf., Reclamation of Contaminated Land*, Eastbourne, SCI, London.
37. Smith, M.A. and Beckett, M.J. (1982) An international study of contaminated land. In: *Proc. Conf., Management of Uncontrolled Hazardous Waste Sites*, Washington DC., HMCRI, Silver Spring, Maryland, 431–433.
38. Smith, M.A. (ed.) (1985) *Contaminated Land*, Final report of the NATO/CCHMS study group, Plenum, New York.
39. Lord, D.W. (1983) The proposed British Standard Code of Practice for the Identification and

Investigation of Contaminated Land. In: *Proc. Conf., Reclamation '83*, Grays, Industrial Seminars Ltd., Tunbridge Wells, 176–184.

40. Cottington, J. (1984) Investigation of contaminated land: Comments on the Draft Code. *Civil Engineering*, February issue, 38–41.
41. Haines, R.C. (1984) Contaminated land-site surveys and hazard assessments. In: *Proc. Conf., Environmental Contamination*, London, CEP Consultants, Edinburgh, 282–287.
42. Mason, D. (1984) Fixing a safe level of contaminant. *Surveyor*, August 2 issue.
43. Tyler, K.J. and Simms, D.L. (1984) Letter to the Editor, *Surveyor*, August 23 issue.
44. Simms, D.L. and Beckett, M.J. (1985) Contaminated land: Risk analysis and safety margins. Paper presented to Conf., Contaminated Soil, Utrecht, Netherlands.
45. Thompson, M.J. and Edmondson, D.A. (1984) Wither derelict land grant, *The Planner*, **70**, 16–18.
46. Anon. (1983) Efficiency in Britain's environmental policy-making: Minis reveal all. Environmental Data Services Report 107, 9–12.
47. Anon. (1984) Staff cuts at DoE halted, but failure to meet targets continues. Environmental Data Services Report 115, 4–5.
48. Royal Commission on Environmental Pollution (1984) Tackling Pollution—Experience and Prospects, RCEP 10th Report, HMSO, London.
49. Royal Commission on Environmental Pollution (1985) Managing Waste: The Duty of Care, RCEP 11th Report, HMSO, London.
50. Hillman, T.C. (1984) The Waste Disposal Authority role. In: *Proc. Conf., Hazardous Waste Disposal and the Re-use of Contaminated Land*, London, SCI, London.
51. Khan, A.Q. (1980) Investigation and treatment of Ravenfield tip. In: *Proc. Conf., Reclamation of Contaminated Land*, Eastbourne, SCI, London, F3/1–F3/11.
52. Cooney, P. (1980) Heavy metal land contamination: Background levels and site case histories in the London Borough of Greenwich. Papers to a Workshop, Case histories of contaminated land—Local Authorities viewpoints, Research note 16, JURUE, Birmingham.
53. Chapman, W.B., Baker, P. and Burns, (1976) Some guidelines on the use of contaminated land. *GLC Materials and Structures Bulletin*, **98**, 1–8.
54. Kelly, R.T. (1980) Site investigation and materials problems. In: *Proc. Conf., Reclamation of Contaminated Land*, Eastbourne, SCI, London, B2/1–B2/14.
55. Anon. (1984) *Proc. Conf., Contaminated Land—the London Experience, London Environmental Supplement*. **7**.
56. Carpenter, R.J., Goaman, H.F., Lowe, G.H. and Pecksen, G.N. (1985) Guidelines for site investigation of contaminated land, *London Environmental Supplement*, **12**, 1–9.
57. Pecksen, G.N. (1985) Methane and the development of derelict land. *London Environmental Supplement*, **13**.
58. Hoomans, J.P. and J.W. Stellingwerff, (ed.), (1982) Operation Lekkerkerk West, *Pt/Civiele Techniek*, **1**, 3–44.
59. Hoogendoorn, D. (1984) Review of the development of remedial action techniques for soil contamination in the Netherlands. In: *Proc. Conf., Management of Uncontrolled Hazardous Waste Sites*, Washington DC., HMCRI, Silver Spring, Maryland, 569–575.
60. Winsemius, P. (1984) Opening address to the Colloquium: Development of soil cleaning techniques, De Reehorst in Ede, Netherlands, Ministry of Housing, Physical Planning and the Environment, The Hague.
61. Anon. (1983) *Handbook of Soil Reconstruction Techniques*, Ministry of Housing, Physical Planning and the Environment, The Hague.
62. deLeer, E.W.B. (1986) Thermal methods developed in the Netherlands for the cleaning of contaminated soil. In: *Proc. Conf., Contaminated Soil*, Utrecht, Netherlands, Martinus Nijhoff, Dordrecht, 645–654.
63. Assink, J.W. (1986) Extractive methods for soil decontamination, a general survey and review of operational treatment installations. *Ibid*, 655–667.
64. Smith, M.A. (1982) NATO/CCMS pilot study on contaminated land—register of important sites, Building Research Establishment, Garston.
65. Stief, K. (1984) Remedial action for groundwater protection: Case studies within the Federal Republic of Germany. In: *Proc. Conf., Management of Uncontrolled Hazardous Waste Sites*, Washington DC., HMCRI, Silver Spring, Maryland, 565–568.
66. Franzius. V. (1986) "Remedial actions on contaminated sites—approaches to solve the

problems of abandoned sites in the Federal Republic of Germany". In: *Proc. Conf., Contaminated Soil*, Utrecht, Netherlands, Martinus Nijhoff, Dordrecht, 269–284.
67. Anderson, J.K. and Hatayama, H.K. (1984) Beneficial reuses of hazardous waste sites in California. In: *Proc. Conf., Management of Uncontrolled Hazardous Waste Sites*, Washington DC., HMCRI, Silver Spring, Maryland, 363–367.
68. Salvesen, R.H. (1984) Downtown carcinogens—a gaslight legacy. *Ibid*, 11–15.
69. Brink, J.M. and Tarlton, S.F. (1984) Cleanup of radium processing residues in a highly urbanised area. *Ibid*, 445–448.
70. Raloff, J. (1980) Disaster on 99th street. *New Scientist*, June 19th issue, 298–300.
71. Smith, R.J. (1982) The risks of living near Love Canal. *Science*, **217**, 808–811.
72. Anon. (1982) Superfund's Remedial Response Program. US Environmental Protection Agency Report, Washington DC.
73. Anon. (1984) National Priorities List, US Environmental Protection Agency Report, Washington DC.
74. Beardsley, T. (1984) Agency pleads for less support. *Nature*, **310**.
75. Thomas, L.M. (1984) EPA fights hazardous waste. *EPA Journal*, **10**, 4–7.
76. Frank, J. (1984) Constructive criticism on the implementation of the Superfund program—a State perspective. In: *Proc. Conf., Management of Uncontrolled Hazardous Waste Sites*, Washington DC. HMCRI, Silver Spring, Maryland, 532–534.
77. Anon. (1984) Groundwater protection standards for hazardous waste disposal facilities: Will they prevent more Superfund sites. Report from the Office of Technology Assessment, United States Congress, Washington.

2. Types of contaminated land

G.D.R. PARRY and R.M. BELL

2.1 Introduction

Throughout this book the reader is reminded that the recognition of contaminated land, as a specific problem of redevelopment, as distinct from general derelict land problems, is relatively new. As a result, the full extent of its occurrence and the forms it can take are therefore, to some degree, still uncertain. A further consequence is the bewildering array of statements and definitions presented in the literature relating to contaminated land. (Chapter 1)

The previous and subsequent chapters do however provide an indication of the size of the problem, the main contaminants that can occur and their effects. A major theme that runs through this book is the complex and diverse nature of contaminated land, its origins and the various treatments needed to facilitate pollution control and allow redevelopment.

If providing a definition for contaminated land acceptable to all interested groups is difficult then the classification of land into 'types' is even more difficult. A number of attempts have been made to classify contaminated land in a variety of ways. In general, all these classifications resort to a separation of sites in terms of their previous industrial use. In strictly adhering to this approach, classification will necessarily preclude sites which have been contaminated in a vicarious manner. It will not, for example, take into account those sites which have become contaminated by industrial activity elsewhere, for example, where pollutant transfer by air or water transportation has occurred.

This chapter identifies the main types of contaminated land, both in terms of the processes involved and former industrial land uses and their associated activities.

2.2 Transportation of contamination

The later sections of this chapter are devoted to the particular associations of specific past industrial uses of land to its likely contamination problems. Chapter 4 stresses how this causative link can help the assessment of contaminant levels and distribution.

However, it is also important to note that contamination can be moved many kilometres from the industrial process that created it by the natural effects of wind and water movements, or by a deliberate off-site tipping of waste products. The actual examples quoted below indicate that such transportation effects can give rise to heavily contaminated soils.

2.2.1 *Atmospheric fallout*

Examples include:

(a) Oxide fume emissions from metal refining works
(b) Fluoride emissions from aluminium refining works
(c) Acidic precipitation from chemical works.

The importance of this process can be indicated by reference to the large scale impacts that resulted from the decades of heavy metal smelting in the Lower Swansea Valley, Wales. This area was the UK centre for non ferrous metal smelting from the early 1700s to the first part of the 1900s. By the 1950s over 300 hectares of land had been completely blighted, physically and chemically, by the tipping of smelter wastes, but this was only a part of the impact of industrial development. Significant areas adjacent to the industrial part of the valley had also become seriously contaminated with heavy metals carried and deposited as air pollutants. The degree of this contamination was such that cattle and horses, grazed on adjacent agricultural land in the 1960s died as a result of lead poisoning caused by ingestion of contaminated vegetation (1).

2.2.2 *Contamination by liquids*

Examples include:

(a) Transport of contaminants by river flow onto flood plains
(b) Spillage and leakage of wastes into the water table
(c) Application of sewage sludge onto land.

A spectacular example of land contamination as a result of transport by streams and rivers is provided by Parc Mine in North Wales, United Kingdom. Parc Mine operated as a lead and zinc source from the 1800s to 1954. A result of this mineral exploitation was the development of a waste rock and fine tailings dam of 250 000 t in a valley head on a tributary of the river Conwy. During a series of storm events, primarily during the winter of 1963–64, erosion of the tailings dam occurred and effectively destroyed 6 ha of agricultural land on the flood plain of the river Conwy more than 1 km from the tailings source, and seriously damaged a further 5 ha in the immediate vicinity. It was estimated that some 13 000 t of material were transported to the flood plain, and that this was equivalent to 107 t of lead, 3 t of cadmium and 419 t of zinc. This dispersal was linked to several instances of lead poisoning in cattle, and zinc toxicity in cereals grown later on the affected area (2).

2.2.3 *Contamination by solid waste disposal*

Examples include:

(a) Mining soil
(b) Industrial slags
(c) Landfill of industrial waste
(d) Uncontrolled waste disposal.

There are, of course, numerous examples of land contamination resulting from solid waste disposal, however, one of the best documented cases arose in Holland. In Dordrecht, the Merwedepolder housing development occurred on a former landfill site. Approximately 7 ha of land was used for waste disposal between 1965 and 1971. The materials tipped on the site included domestic refuse and chemical waste from a phenol producing firm.

The site was covered with varying depths of river sediments and levelled in 1971. Subsequently, housing development took place. This development altered the topography of the site, bringing contaminated material to the surface. In 1981 an extensive investigation began following complaints by residents of persistent organic chemical odours. Exposure of the foundations revealed solid and tarry wastes containing benzoic acid, phenol, diphenyl ether and other complex organic substances.

Benzene, trichlorethane and biphenyl ether and other organic vapours were found in low concentrations in the air in a number of the houses. Although no health effects were proved by the Authorities, public reaction resulted in 106 houses being demolished in one quarter of the site and extensive groundworks undertaken to isolate the pollutants from the remaining residential structures (3).

2.3 Main sources of land contamination

In considering sources of contamination, certain classes of substances are associated with specific industries or activities, regardless of their location, and are commonly found in land associated with that industry. The main industries and activities which later create contaminated land problems may be listed as follows:

(a) *Mineral extraction and subsequent landfill*
Quarries
Gravel pits
Coal (deep mining and opencast)
Clay pits

(b) *Metalliferous mining (ores)*
Iron
Tin
Lead
Copper
Tin

(c) *Process works*
Chemical (petrochemical)
,, (organic)
,, (inorganic)
Steelworks
Gasworks
Power stations
Smelters
Tanneries
Textiles

(d) *Transportation*
Ports
Canals
Railways
Roads
Airports

(e) *Bulk wastes*
Pulverized fuel ash (pfa)
Industrial refuse
Town wastes
Sewage works
Lime waste (soda-ash)
Chromate waste
Spent oxide
Radioactive wastes

The above list is not exhaustive, but a detailed examination of specific examples provides an indication of the range of industries that contribute to the problem of soil contamination.

2.3.1 *Non-ferrous metal mining and processing sites*

Non-ferrous metal mining is one of Britain's older industries though few operational mines exist today. The relics of the mines and processing plant still exist in the major ore fields of the United Kingdom, in Devon, Cornwall, the Mendips, north and south Pennines, mid and north Wales, and parts of Ayrshire.

Many of the older sites were heavily contaminated by metals in the past, since crude extraction techniques allowed tailings to be produced as a waste product which contained up to 10% metal. Modern flotation techniques, however, reduce metal concentrations in tailings to 0.1% (4).

The major contaminants of such sites are normally the metals that were actually mined, for example lead, zinc, copper and tin, or the associated metals that occurred within the mineralized deposit. For example, arsenic often naturally occurs with tin, while cadmium usually accompanies lead and zinc. The impact of these contaminants is often quite striking in terms of vegetation toxicity and hazards to livestock.

2.3.2 *Sewage works and farms*

Sewage works and farms range in size from small to large treatment works which, with their associated effluent and sludge disposal areas and farmland, may cover many hectares. Thus the scale of the works installations and the extent of any contamination are likely to vary considerably from site to site. Possible hazards to animal health and toxicity to plants may also be encountered on agricultural or horticultural land which has been treated by the repeated application of sewage sludge or effluent for many years.

Sewage sludges can contain a wide range of metal contaminants which are related to the various types of industrial sewage input. Common contaminants include lead, zinc, copper, cadmium and nickel whilst almost any metal, including arsenic, boron, mercury and manganese can occur less frequently and at lower concentrations (5).

2.3.3 *Scrapyards*

Sites used for breaking up of redundant or obsolete manufactured items such as vehicles, electrical equipment, machinery, etc. are among the most commonly encountered examples of contaminated land. The nature and degree of contamination depends upon the activities carried out on the site concerned, but can commonly include cyanides, sulphates, acids and alkalis and various organic compounds. The presence of these potentially carcinogenic pollutants in the ground is often more important than the presence of metals which will also be found in high concentrations.

Contaminants can be present in a variety of forms including liquid and solid wastes and sludges. It is usually impossible to generalize over the contaminant distribution on the site, since the ground surface will frequently be covered with metal dust, waste oils and other organic contaminants. Common metallic contaminants are lead, zinc, copper, cadmium and nickel with other metals originating from the types of items broken up on site (6).

2.3.4 *Railway land*

Redundant railway land and sidings occur in all parts of the United Kingdom and most are not contaminated. In some areas, however, the trackways have been constructed of waste ash and clinker from local industrial sources and may thereby be contaminated. Common contaminants include lead, zinc and cadmium with high concentrations of copper, nickel, and manganese. Chromium and arsenic occur less commonly, as do contaminants similar to those found on gasworks sites.

2.3.5 *Iron and steelworks sites*

After the nationalization of these industries in 1947, in the United Kingdom many small works were amalgamated into much larger units, covering perhaps areas of 300–600 ha. Much of this land is now covered by iron, steel and blastfurnace slags containing elevated concentrations of manganese, boron, chromium, lead, zinc and cadmium. The actual metallic content of the slag depends on the type of iron or steel being produced, with the slags resulting from 'special steel' production containing a wider range of metals than normal (7).

As the metals are usually fused into the slag, their impact is dependent upon the likelihood of the slag being weathered and breaking down to release the metals into the environment. This does not often occur and the impact of such wastes is generally low.

Many iron and steelworks, however, and their own energy-producing plants, for example, the Consett Steelworks, United Kingdom, included a coal carbonization plant, while the Scunthorpe Steelworks, United Kingdom, included a coal-fired power station. The potential contaminants arising from these plants are discussed below in the section on gasworks.

2.3.6 *Dyeworks*

The petrochemical industries supply the primary materials for dyestuffs manufacture. These include many organic acids and solvents which are converted into dyestuff intermediates by reactions with dichromates, lead, nitric and sulphuric acids and some reactive gases. Dyestuffs are then produced by further chemical treatment of the intermediates.

The chemicals used in dyeworks are principally organic compounds and many are potentially carcinogenic. It is their presence which constitutes the greatest contaminant threat, rather than the presence of metals. If, however, the dyestuffs were utilized on site to dye materials, then common mordants used to make fast the dyes may also be present on site. These include chlorates, acetates, and sulphides of barium, chromium, copper, lead and nickel. Dyework sites are generally accepted as amongst the more difficult ones to redevelop safely.

2.3.7 *Coal carbonization on gasworks sites*

The production of coal gas and coke involved heating coal in an oxygen-deficient atmosphere which produced a variety of reactions resulting in a mixture of gaseous and liquid products. It is ground contamination by these products, and others arising from secondary processing and by-product manufacture that causes serious problems in redeveloping such land (8).

The major contaminants likely to be found are coal tars, phenols and spent oxide. The latter, which was used as a purifying agent for coal gas, contains free and complex cyanides, elemental sulphur, sulphate and sulphide. Invariably, the degree of contamination will be high but it is unlikely to be evenly distributed all over the site.

Pockets of contamination tend to occur in areas of the sites associated with former production and storage areas, for example, purifier sheds and tar wells. Isolated surface contamination may also occur in areas used for waste disposal of spent oxide and lime wastes used in the purification process.

The most significant contaminants in terms of redevelopment are coal tars and sulphates. The major problem associated with the former is the risk of prolonged skin contact by children playing on the site, or by workers during redevelopment, which can give rise to carcinogenic effects. The generally high sulphate concentrations in the ground on such sites will usually require sulphate-resistant cement to be used in structures. Additionally, high concentrations of sulphate are detrimental to plant growth and landscaping schemes (9). Gasworks produced vast quantities of waste products, often for periods of more than a century, and so such sites (Chapter 6) are usually mantled by a thick cover of contaminated wastes.

2.3.8 *Coal-fired power stations*

The main product of modern coal-fired powers stations is the large quantity of pulverized fuel ash (pfa) produced during operation. This material generally has a fine uniform particle size and is disposed of to lagoons. Redevelopment of the pfa disposal areas must take into account the extremely high pH often exhibited by the ash (pH 11–12) and its high concentrations of arsenic and boron.

Redevelopment of the site of a power station itself can pose problems in addition to those of the pfa disposal sites. In common with older solid fuel utilization structures, there is the likelihood of residues of asbestos insulation remaining on the site following demolition. In addition, residues of fuel remaining in the coal stock yards often give rise to problems of combustibility.

2.3.9 *Landfill sites*

The economics of solid waste disposal are such that landfill will remain the favoured disposal method in the United Kingdom for some considerable time. Since the Control of Pollution Act (1974) the disposal of all solids wastes in the UK has been controlled and records of materials disposed into landfill are now maintained. For sites completed before this time there is however no record and it can be expected that a wide range of contamination will have occurred.

The main problem of such sites is of gas production from the biological breakdown of the deposited wastes. Gases produced include carbon dioxide, hydrogen sulphide and methane, which not only inhibit the growth of vegetation, but, under certain conditions, can migrate into buildings and may accumulate to explosive concentrations (see Chapter 11).

If all wastes are fully broken down and gas production has ceased, then the presence of metal contaminants which usually remain becomes more important. The actual metals likely to be present depend on the nature of the tipped wastes but will include lead, zinc, copper and nickel as well as less common ones.

Additional problems of redevelopment of these sites should also be considered. Tip fires are recognized as a hazard during the operational phase of landfill, but can also occur during the redevelopment phase. Exposure of combustible material to the air can give rise to spontaneous combustion, but the most likely cause is considered to be direct ignition or heating from an external source.

Further problems can arise during and after redevelopment due to the generally unstable nature of the ground, and the potential presence of voids caused by subterranean fires which in large tips may have continued for years (10).

2.3.10 *Pharmaceutical manufacturing sites*

Pharmaceutical manufacturing including toiletries and cosmetics, spans a wide range of products, including medicines, dental speciality materials, hormones and vitamins for incorporation into foodstuffs as well as toothpaste, hair sprays, deodorants, shampoos and perfumes.

The vast majority of wastes from this manufacturing are liquid effluents which are discharged to rivers and estuaries or piped to coastal waters or sewers. The majority of solid wastes are disposed of in landfill.

Practically every commercially available organic solvent is used in pharmaceutical manufacturing, and all could technically be present on land occupied by such an industry. Thus the redevelopment of such sites should be undertaken with especial care and attention.

2.3.11 *Tanning and fellmongering*

There is no single process for the production of leather and as many as fifty different chemical or mechanical processes may be involved. A wide range of products necessitates the use of many process chemicals and the industry is characterized by batch processing in aqueous media and the incomplete exhaustion of process chemicals. Leather is essentially animal skin protein combined with tars, oils, dyes and other finishing agents. The range of potential wastes is large and includes both organic and inorganic compounds including white spirit and kerosene used for skin de-greasing, and chromium sulphate for tanning.

The presence of organic chemical contaminants poses the greatest problems in redevelopment of sites with this former use, although contamination by metals such as chromium may pose additional problems.

2.4 Conclusions

Land is an invaluable commodity which will always be subject to redevelopment as the industrial society and processes change. It is important to appreciate that land available for redevelopment may have been contaminated as a result of its previous use, or by the transport and deposition of pollutants from other sites. In considering the 'types' of contaminated land it is necessary to take into account both these processes which interrelate in a complex manner. In general, however, as indicated in this chapter, it is the previous use of a site which is most commonly related to its contamination and so forms an important guide in assessing whether or not a site is likely to be contaminated.

References

1. Bromley, R.D.F. and Humphrys, G. (eds.) (1979) Dealing with dereliction. The redevelopment of the Lower Swansea Valley. University College of Swansea.
2. Welsh Development Agency (1977) Rehabilitation of Metalliferous Workings at Parc Mine nr. Llanrwst. A baseline study of Environmental Conditions.
3. Bins-Hoefnagels, I.M.G. and Molenkamp, G.C. (1985) Case studies of soil pollution in some Dutch urban areas. *Proceedings of the First International TNO Conference on Contaminated Soils*. 11–15 November Utrecht, The Netherlands, 769–779.
4. Johnson, M.S. and Bradshaw, A.D.B. (1977) Prevention of Heavy Metal Pollution from mine wastes by vegetative stabilisation. *Trans. Inst. Min. and Metal A.*, **86**, 47–55.
5. Interdepartmental Committee on the Redevelopment of Contaminated Land (1983) Notes on the redevelopment of sewage works and farms. ICRCL 23/79. Department of the Environment.
6. Interdepartmental Committee on the Redevelopment of Contaminated Land (1980) Notes on the redevelopment of scrap yards and similar sites. ICRCL 42/80. Department of the Environment.
7. Smith, M.A. (ed.) (1985) *Contaminated Land Reclamation and Treatment*. NATO Challenges of Mod. Soc. Vol. 8, Plenum, New York.
8. Interdepartmental Committee on the Reclamation of Contaminated Land (1979) Notes on the redevelopment of gasworks sites ICRCL 18/79, Department of the Environment.
9. Wilson, D.C. and Stephens, G. (1981) Problems arising from the redevelopment of gas works and similar sites. United Kingdom Atomic Energy Authority, Harwell. AERE-R 10366.
10. Smith, M.A. (1982) Methane and development. Building Research Establishment note 48/82, Department of the Environment.

3 Main types of contaminants

R. C. HAINES and MARY R. HARRIS

3.1 Introduction

It will be appreciated from the preceding chapters that because of the inherent diversity in the materials and processes used in industrial activities it is inevitable that many different types of contaminated land have emerged. Furthermore, there is a corresponding variability in the range of hazard that particular contaminants are considered to pose, either to the environment or public health.

The purpose of this chapter is to provide some detailed background information on particular contaminants which are both representative of a group of contaminants and which are commonly encountered in practice. Information of this type is useful for a better understanding of the reasons why a particular substance, by virtue of its physical, chemical and/or biological particular contaminants are considered to pose, either to the environment or public health.

It is intended to concentrate on those groups of contaminants for which UK guidelines for the redevelopment of contaminated land are currently available. This is primarily so that information here can be used to complement current guidelines, hopefully in such a way that the utility of guideline levels is enhanced. Secondly, for some substances, although there are good reasons to suppose that the presence of the material in soils would constitute some degree of hazard, there are insufficient data to indicate the significance of the hazard in the context of contaminated land. It is therefore impossible at this point in time to provide comprehensive information on more than a very limited number of potential contaminants.

In spite of these difficulties, it has been felt necessary to mention briefly some relatively unfamiliar contaminants either because concern regarding them has been expressed in Europe or North America as a result of particular circumstance or policy, or because prospective changes in industrial structure suggests that these materials may well become the familiar soil contaminants of the future.

The chapter therefore provides a state of the art review of the knowledge concerning the sources, toxic and/or other undesirable properties and current guideline information for eleven specific contaminants and general information on a small number of other important groups of potentially hazardous materials.

3.2 Metallic contaminants

3.2.1 *Arsenic*

Source of contamination

Soil contaminated with arsenic has arisen as a result of the mining and smelting of the metal, its extensive use in agricultural preparations (such as animal dips, pesticides) and its presence in wood preserving products. The table below shows the various uses, and potential sources of arsenic in the environment.

Insecticides e.g. lead arsenate, calcium arsenate
Weed killers e.g. sodium arsenate
Cattle dip e.g. arsenic trioxide
feed additives e.g. arsanilic acid
Wood preservatives e.g. arsenic trioxide

In the manufacture of:

Glass	Anti-fouling paints
Pigments	Textiles
Enamels	Fireworks

Levels of arsenic in soils

Arsenic concentrations in soils average 5–6 mg/kg, but can vary considerably according to region with some Dartmoor soils reaching 250 mg/kg*. Contamination of soils associated with the industries and materials described above can also substantially elevate arsenic soil concentrations above the 5–6 mg/kg level as indicated below:

Soil concentration As (mg/kg)	Source of contamination	References
up to 2500	Mining/smelting	2
21–42	Pesticides use	3
9–58	Wood preserving	4

Arsenic in topsoils is exposed to atmospheric oxygen and is usually present in the pentavalent form, that is as arsenates, with H_2AsO_4 prevailing in acidic conditions and $HAsO_4$ in alkaline conditions. However, under reducing conditions, arsenite is likely to become the dominant form in addition to being the form in which most industrial contamination takes

*DoE ICRCL trigger concentration—see Chapters 1 and 4.

place. Bacterial action can bring about the oxidation of arsenite to arsenate, and other reactions can take place in the soil which methylate the element, methylation generally reducing toxicity.

Principal hazards

In the industrial environment two forms of arsenical poisoning can occur; acute poisoning arising as a result of the ingestion of (20 mg) arsenic or by inhalation of toxic arsine gas. Chronic poisoning results from long-term exposure to, and accumulation of the metalloid arising from the gradual absorption of arsenical compounds either by inhalation or ingestion. Arsenic-containing dust can also behave as a skin irritant causing inflammation and ulceration while arsenic and its compounds are also considered to be carcinogenic.

The toxicity of arsenic to plants depends on oxidation state and form of the element, with the arsenite form being more toxic than the arsenate form. Plants have been shown to be tolerant to high levels of arsenic in soils contaminated as a result of mining operations (5). Extremely high levels of the metal can occur in metal tolerant plants (e.g. 3470 mg/kg dry matter in one sample of *Agrostis tenuis*) although for plants in general, growth reduction is likely to take place before levels of the element in plants reach toxic levels. Barriers to the trans- location of the metal from root to shoot are thought to exist (1) and the poisoning of livestock by arsenic is thought, in some cases, to be due more to the ingestion of arsenic rich soils than contaminated herbage (2).

Arsenic chemistry in aquatic systems is complex with four stable oxidation states (+5, +3, 0, −3) thought possible. In oxidized surface waters arsenate species are stable, with arsenite or even arsine becoming predominant if reducing conditions develop. Water pollution by arsenic is a possibility in some circumstances, for example, the combustion of timbers treated with arsenic containing preservatives can give rise to an ash containing high concentrations (up to 5%) of arsenic, a proportion of which may be water soluble.

The potential hazards associated with arsenic found on contaminated sites are therefore related primarily to the possibilities of inhalation, ingestion and contact with arsenic rich soils and dusts and also to the possibilities of phytotoxicity, plant accumulation of the metalloid and water pollution.

Guideline levels

Planned uses	Trigger concentration (mg/kg) air-dried soil
Domestic gardens, allotments	10
Parks, playing fields, open space	40

3.2.2 *Cadmium*

Sources of contamination

The environmental significance of cadmium has been reviewed comparatively recently by the Department of the Environment in the form of a Pollution Paper (6). This report deals with the production and sources of cadmium in the environment as well as examining the various pathways of cadmium intake into man. The main uses and sources of cadmium are briefly summarized below:

Mining and smelting (secondary product of other non-ferrous metals)
Pigments (e.g. plastics, paints)
electro-plating
batteries, photo cells and solar collection
alloys and solders
polyvinylchloride stabilizer
Fungicides and anti-helminthics

Soil contamination by cadmium has occurred as a result of emissions from non-ferrous smelting, particularly of zinc, but also of lead, in association with primary iron and steel production and as losses from mining sites. Cadmium is also present in street dusts in association with zinc (from car tyres containing zinc oxide) and lead (from petrol additives). Fertilizers containing relatively high concentrations of cadmium may contribute to the contamination of agricultural land (7) as may the application of contaminated sewage sludge to land (8). A further important source of cadmium in soils is the disposal of both industrial and domestic wastes containing cadmium-bearing items. It has been suggested that in the UK landfilling is, and will remain, the major outlet for cadmium-bearing wastes with the quantity of material requiring disposal set to increase. For example, cadmium from just one source, nickel-cadmium batteries, is set to rise from its current level of 140 tonnes per year to about 300 tonnes per year by the late 1980s (9).

The amount of cadmium in soils is highly dependent on the underlying rock but is present in most soils at concentrations of 1 mg/kg or less. Local enrichment in some soils can occur as a result of mineralization of parent rock type (2). These and other sources of cadmium can considerably elevate the level of the metal in soils as shown below:

Soil concentrations Cd (mg/kg)	Source of contamination	Reference
up to 30	Mineralization (Derbyshire)	2
2–336	Zn mining (Shipham)	10
1.5–64	Sewage amendment	11
1–17	Urban garden soils (national)	12
1–40	Urban garden soils (London)	12
0–50	Former waste disposal site	13

The origin of the contamination, as well as the form in which the element is present in the soil are important factors governing the transfer of the metal from the soil to humans via the most important dietary route (see below).

Principal hazards

Cadmium has no known biological function, but it is readily accumulated in man, preferentially locating in the soft tissues, particularly the liver and kidney. The metal can exert a number of detrimental effects on health with long-term low-level exposure to the metal being of prime concern. Under these circumstances of exposure, health effects have been stated (14) to include:

(i) Kidney damage leading to increased excretion of protein, amino acids, glucose, phosphate and calcium
(ii) Disturbance to the metabolism of other metallic species in the body, including iron (leading possibly to anaemia), zinc and copper (copper deficiency has been implicated in the development of emphysema)
(iii) Long-term effects are thought to include the development of hypertension and cancer of the prostate.

Cadmium is absorbed by man both via the respiratory and alimentary system, with food and to a lesser extent drink, being the most important routes for non-occupationally exposed adults. Cigarette smoking is also an important route of exposure.

Although most of the UK population is considered to take in less than the 400–500 μg Cd (provisional maximum tolerable weekly intake from all sources suggested by the United Nations Food and Agricultural Organisation/World Health Organisation Committee on Food Additives), some sections of the community are considered to regularly exceed this amount (15). There are, however, difficulties in establishing a direct link between marginally excessive intakes of the metal with significant health effects. Recent studies at Shipham for example have shown that although residents of the village, located in a former zinc mining area, do have higher than normal intakes of the metal, no significant adverse health effects have been detected (10). However, the potential magnitude of exposure at Shipham, coupled with uncertainties associated with confounding effects and long lead time, have led to the general view that intakes of cadmium via dietary sources must be minimized.

In the context of the redevelopment of contaminated land, the principal hazard associated with the metal is the possibility of long-term ingestion of contaminated food and water by vulnerable groups, principally children.

The uptake of cadmium by plants depends on a number of factors including pH (lower pH generally increases cadmium availability) binding capacity of the soil (lower availability in soils with, for example, a high cation exchange capacity, or organic matter content), texture and source of contamination.

Recent work (11) has shown that in both 'inorganically' (e.g. mining wastes) polluted and 'organically' (e.g. sewage amendment) polluted soils, cadmium was present in the soil solution primarily in the cationic (Cd^{2+}) form. Differences in plant species, or even cultivar of the same species, can account for differential uptake of the metal grown in the same soil (16, 17). In general, it is accepted that when considering plants for human consumption, leafy vegetables such as lettuce or spinach are likely to take up more metal than vegetables grown primarily for their root or tuber e.g. parsnip, potato, although it should be remembered that in some circumstances, the contribution of cadmium to the diet from root crops can become significant because of the large proportion of root crops, compared to other vegetables, in average diets (18).

There is some evidence both of phytotoxic effects of cadmium at high levels of cadmium exposure and tolerance to high soil levels of the metal (19).

Guideline levels

Planned uses	Trigger concentration (mg/kg) air dried soil
Domestic gardens, allotments	3
Parks, playing fields, open space	15

3.2.3 *Lead*

Sources of contamination

Soil contamination with lead and its compounds has been documented in relation to the primary and secondary smelting of the metal, mining activities, the use of lead as an anti-knock agent in petrol and as an additive in paints and pesticides. Severe local contamination of soils has arisen as a result of car-breaking activities and battery recycling carried out on non-commercial premises, while the application of contaminated sewage sludge can also increase soil levels of lead.

The various sources and uses of lead can be summarised as:

Mining and smelting	lead compounds in petrol
Plating	pigments for paints and glazes
Batteries, cables, foils	glass manufacture

Levels of lead in soils

The normal level of lead in British soils is generally quoted at between 2–200 mg/kg. Contamination resulting from activities and materials described above can however result in much higher levels of the metal in soils as indicated below:

Soil concentration Pb mg/kg	Source of contamination	Reference
10 000 or more	Mineralization	2
up to 51 000	Secondary smelting (Canada)	20
up to 40 000	Secondary smelting (UK)	21
13–14 125	Urban garden soils (national)	12
1–13 680	Urban garden soils (London)	12
greater than 600	Areas of former waste disposal site with carbreaking	22
31–938	Sewage amendment	11

Lead is generally considered to be comparatively immobile in soils with much of the material deposited on the surface, due to atmospheric deposition, remaining fixed in the upper layers of the soil. Most lead in soils is thought to occur as Pb^{2+} in relatively insoluble compounds and associations. However, more soluble compounds e.g. chlorides, bromides and sulphates may be released into the environment as a result of human activities and lead derived from paint sources (carbonates, oxides, chromates) is more soluble than most of the lead present in soil).

Principal hazards

Detrimental effects associated with lead have been recognized for a long time. Clinical symptoms associated with lead poisoning include anaemia, various digestive disorders, and central nervous system (CNS) effects. However, present concern about the metal relates to possible deleterious effects of lead which may take place at low levels of exposure. Sub-clinical lead poisoning occurs when small amounts of lead are taken in over a prolonged period of time. Biochemical processes in the body are affected and these disturbances are detectable long before the symptoms of clinical lead poisoning become apparent. There is some controversy over whether this type of exposure to lead is related to various behavioural disorders in children, level of IQ achievement and hyperactivity. Exposure to lead in drinking water has also been implicated in the incidence of cardiovascular disease in soft water areas. Current policy in the UK is that although there is no conclusive evidence of significant deleterious health effects in individuals with blood lead levels below 35 mg Pb/dl, in order to provide a margin of error, any individual, but children in particular, found to have a blood lead level of 25 mg/dl or over is required to have possible exposure routes checked and steps taken to reduce intakes.

The chief concern with lead contaminated soils and dust rests therefore, on the possibility of a close association between soils and children, in particular, leading to enhanced intake of the metal into the body and the possible adverse health effects outlined above.

The relative immobility of lead, in terms of translocation from the soil to inner or aerial parts of plants is well documented (23) and it is accepted that where lead does become a significant contaminant of food the cause is generally superficial deposits or inadequate washing of vegetables. Although the levels of lead in drinking water can be a significant contribution to intake of the metal if other sources are low, the chief route is considered to be hand-to-mouth transfer of lead from unwashed hands, working surfaces, etc. and from improperly cleaned or washed vegetables. The extent to which hand-to-mouth transfer and thus lead levels in soils and dust relate, in quantitative terms, to lead intakes is difficult to determine, although the relationship between blood lead levels and levels in soils has recently been investigated and defined (24).

Guideline levels

Planned uses	Trigger concentrations (mg/kg) air-dried soil
Domestic gardens, allotments	500
Parks, playing fields, open space	2000

3.2.4 *Copper, nickel and zinc*

Sources of contamination

Soil contamination by the metals copper, nickel and zinc has been observed in connection with mining, smelting and refining operations, the application to land of contaminated sludges and animal slurries, the use of pesticides and the disposal of both industrial and domestic wastes. The table below shows the principal sources and uses of the three metals:

Copper	Nickel	Zinc
Mining	Mining	Mining
Smelting	Smelting	Smelting
Plating	Plating	Plating
Alloys	Alloys	Alloys
Pigments	Batteries	Catalysts
Feedstock additives	Catalysts	Tyres
Pesticides	Ceramics	Feedstock additives

Levels of contaminants in soils

Normal levels of those metals in soils have been stated (2) as:

Copper	2–60	mg/kg
Nickel	2–100	mg/kg
Zinc	25–200	mg/kg

Nickel levels in soils derived from some ultrabasic rocks such as serpentine can contain higher levels than indicated, while mineralization of soils or additions of metals from industry or waste disposal activities can result in much higher levels in soils as indicated below.

Metal	Concentration in soil mg/kg	Source of contamination	Reference
Cu	up to 2000	Mineralization	2
	60–380	Pesticide use	25
	40–730	Sewage amendment	26
Ni	up to 8000	Ultrabasic rocks (UK)	2
	up to 25 000	Nickel refining (Canada)	27
	46–180	Sewage amendment	26
Zn	10 000 or more	Mineralization	2
	250–1800	Sewage amendment	26

Copper commonly occurs in soils in the form of sulphates, sulphides and carbonates of varying chemical complexity and in association with soil organic matter. An important property of the element is that it shows a strong affinity for the organic phase. Zinc can form soluble salts with common soil anions and may also be present complexed to simple organic molecules (28). Some of the metal present in soils is likely to be bound to high molecular weight humic substances. In all three cases, a low pH and a lack of soil binding sites enhances the availability of metals for uptake.

Principal hazards

Both copper and zinc, and possibly nickel, are essential trace elements for both animals and plants and in circumstances of deficiency serious effects on the growth and well being of plants and animals can be observed. However, in the context of contaminated land redevelopment, excessive levels of these metals are considered problematic, primarily in connection with toxicity effects on plant growth (phytotoxicity).

Effects are most marked when low pH and a lack of soil binding sites combine to enhance the availability of the metals to sensitive plant systems. While plants differ in their sensitivity to metals in the soil solution a system has been in use in the UK for some time by which the phytotoxic potential of soils in relation to copper, nickel and zinc can be judged. The measure, known as 'zinc equivalent', was developed for use on soils in which copper, nickel and zinc could be introduced as a result of sewage sludge application and it assumes that copper is twice, and nickel eight times as toxic as zinc; that the toxicities of the three metals are additive; and that soils are maintained to a pH of 6.5 or above. While the concept does not take account of the influence of some important soil and plant factors, it does appear to have some practical relevance in determining the level of sludge which can be added to soils without phytotoxic effects becoming significant and current DoE guidelines for contaminated soils are based on the zinc equivalent concept.

Nickel differs from copper and zinc in that when present as a metallic dust, or as nickel salts, some deleterious effects on human health can develop after direct contact. Thus, direct contact with nickel or solutions of nickel salts can be a common cause of chronic dermatitis in humans and there is some evidence of increased levels of lung and nasal cavity cancers in workmen employed in nickel smelters and refineries. Evidence for this relationship, however, refers to extremely high historical industrial exposures and there is only limited evidence for a link between these health conditions and occupational exposure in other nickel-using industries. Therefore, although there may be some risk of skin irritation amongst individuals on redevelopment work of sites in which extensive nickel contamination is present, the hazard is essentially a short-term one with any long-term risks likely to be minimized as a result of redevelopment.

Guideline levels

Metal*	Planned uses**	Trigger concentrations mg/kg air dried soil
'Available' Cu	Any uses where plants are to be grown	50
'Available' Ni	Any uses where plants are to be grown	20
'Available' Zn	Any uses where plants are to be grown	130

*(i) 'Available' metals to be determined by a standard ADAS method (extractable by 0.5 m EDTA)
(ii) Trigger concentrations apply to the 'worst case' of an acid, sandy soil. In soils of a neutral or alkaline nature phytotoxic effects are unlikely at these concentrations
**(i) Soil pH is assumed to be about 6.5 and should be maintained at this value. If the pH falls, the toxic effects and the uptake of the elements will be increased
(ii) Grass is more resistant to phytotoxic effects than are most other plants and its growth may not be adversely affected at these concentrations.

3.3. Inorganic contaminants

3.3.1 *Cyanides*

Sources of contamination

Although cyanide is a metabolite of some living organisms its presence in soils at excessive levels is usually a result of human activity. Soils contaminated with cyanide can result from waste disposal, for example, of plating bath wastes or other apparatus and materials contaminated as a result of the heat treatment of metals; the dismantling of plating shops where residual materials or contaminated building fabric is present; on sites which have been occupied by former gasworks.

Levels of contamination

The table given below shows the levels of cyanide found to occur under various conditions of man-made contamination.

Soil concentration cyanide (mg/kg)	Source of contamination	Reference
0–4000 (total)*	Former landfill	13
Up to 8000 (total)*	Former gasworks	29
Up to 64 (free)*	Former gasworks	29

* Cyanides can be present in either free or complex form, with different implications for the nature of the hazard associated with contamination

The safe disposal of heat treatment cyanide wastes is discussed in a DoE Waste Management Paper (No. 8) (30), and the document also covers sensible precautions to be taken during the dismantling of premises where cyanides have been handled. One of the most important precautions is that waste cyanides should not be deposited in the same area as waste acids since these can rapidly cause the release of toxic hydrogen cyanide. The release of the compound occurs slowly in any event under landfill conditions because of the acidic action of weak carbonic acid also produced in landfills. Hydrogen cyanide released in this way can diffuse away or undergo hydrolysis to form ammonium formate. Although the process is slow it is considered to be an important degradation reaction along with other biodegradation processes. Despite reactions of this type the persistence of cyanide in landfill is thought to be long-term, particularly if the material is stored in drums and there have been some cases of ground and surface water contamination from cyanide (30).

Cyanide contamination associated with former gasworks arises from the materials used to purify coal gases of sulphur and other impurities. Early processes used slaked lime to remove H_2S, CO_2 and HCN. This material was later replaced by hydrated iron-oxide ('bog iron') because this was capable of regeneration and re-use. Regeneration of the oxide was carried out by spreading the material in thin layers with periodic turning, allowing the formulation of elemental sulphur and the re-use of the oxide. However, material having undergone a number of regeneration steps was considered 'spent' and would be either used as a raw material in the manufacture of sulphuric acid or could be deposited on site, or off site, as waste, subsequently becoming a source of both free and complexed (i.e. ferri/ferrocyanides or thiocyanate) forms.

Principal hazards

HCN and the simple ionic cyanides are amongst the best-known poisons. The cyanide ion is a non-specific enzyme inhibitor, but exerts its powerful toxic effects by inhibiting the enzyme cytochrome oxidase, and thus preventing the uptake of oxygen by living tissues. Rapid death may follow exposure to cyanides as a result the inhibition of cellular respiration.

A single 100 mg dose of sodium cyanide can be fatal if taken orally while 57 mg of potassium cyanide represents a fatal dose for a small child of 20 kg weight. The acceptable daily intake of cyanide is 0.05 mg/kg body weight.

Exposure to a concentration of 300 ppm hydrogen cyanide gas in air is immediately fatal, and concentrated aqueous HCN is rapidly absorbed through skin, eyes and mucous membranes.

The complex cyanides do not represent an acute toxicity hazard (fatal dose for a child is about 40 g) as, although dissociation to form HCN can occur, the process is slow.

However, complex cyanides do appear to exert phytotoxic effects (32) which have been attributed to the acidic nature of spent oxide in addition to other toxic agents present in the material.

Cyanide is toxic to aquatic species (for trout concentrations < 0.15 mg/l) while a further consideration on former gasworks sites is that highly toxic HCN can be evolved if accidental or deliberate lighting of fires occurs on site.

It will be seen from the above that cyanides are considered to present hazard in respect of the possible ingestion of contaminated soils, particularly by children and in respect of phytotoxic effects and possible short-term hazards relating to the lighting of fires.

Guideline levels

Contaminant	Planned uses	Trigger concentration mg/kg air dried soil
Free cyanide	Domestic gardens, amenity area open space	50
	Industrial (no landscaping)	500
Complex cyanide	Any uses where plants are to be grown	50

3.3.2 *Sulphates*

Sources of contamination

Various anthropogenic activities result in the contamination of soils with sulphur and its compounds ranging from acid rain to the dumping of sulphur containing wastes.

Levels of sulphate in soils

The table given below shows some typical levels of sulphate present on sites contaminated as a result of coal processing and waste disposal.

Soil concentration SO_3-(%)	Source of contamination	Reference
Up to 20	Former gasworks sites	29
0.03–20	Former waste disposal site	33

On gasworks sites the most common sulphate salts likely to occur are ammonium ferrous sulphates. However, calcium sulphate residues may also occur as a result of the historical use of the lime to purify coal gases.

Principal hazards

Sulphates can represent a toxicity hazard with ferrous sulphates being more toxic than ammonium sulphate. For ferrous sulphate the ingestion of 7.8 g (or about one mouthful) has been proved fatal for a small child. Sulphates present at excessive concentrations ($>$400 mg/l) are not considered acceptable in drinking water supplies, particularly in the presence of magnesium and sodium ions because of the possibility of gastro-intestinal irritation.

Sulphates also exert some phytotoxic effect although plants differ in their ability to withstand high levels of sulphate in the soil solution. The problem of phytoxicity is marked in acid soils. Soil sulphate levels in excess of 200–300 mg/kg are considered to be of concern in relation to plant growth (29).

One of the most pressing problems associated with high levels of sulphates in the soil involves the durability of building materials when placed in chemically hostile environments. Both concrete and cast iron piping are considered vulnerable to elevated sulphate concentrations and special precautions such as the use of sulphate-resisting cements and protective coatings for piping may be considered necessary if sulphate levels in excess of 1000 mg/kg in the soil or 300 mg/l in water are present on site (29). It should be remembered that it is the level of soluble sulphate which is of the greatest concern with respect to the durability of building materials.

The microbial transformation of sulphates to sulphide taking place in anaerobic waterlogged soils is potentially an additional undesirable effect of sulphate contamination. Not only does the reaction lead to the formation of highly toxic hydrogen sulphide with its associated problems of odour and plant growth effects, but also toxic sulphide salts may be formed.

The principal hazards associated with sulphate contamination in soils can therefore be summarized to include:

Materials attack
Ingestion hazards and water pollution
Phytotoxicity problems

Guideline levels

Planned uses	Trigger concentration mg/kg air dried soil
Residential, commercial and industrial developments with no landscaping	1000
All other uses including domestic gardens, open spaces and landscape areas	200

3.4 Organic contaminants

3.4.1 *Phenols*

Source of contamination

Monohydric and polyhydric phenols are by-products of the coal carbonization industry and are also used extensively as feedstock in the petroleum based chemical industries. The term phenol can be used for any aromatic compound whose molecule contains one or more hydroxyl (OH) group although phenol is also the name given to the simplest of these compounds, C_6H_6OH.

Level of phenol in soils

Industrial production of coal gas and the disposal of mixed industrial wastes can cause higher levels of phenol to be present in soils as indicated below.

Soil concentraion phenols (mg/kg)	Source of contamination	References
Up to maximum 1800	Former gasworks sites	29
<0.5–26	Former landfill site	34

Phenol is not considered to bind effectively to soil particles and being readily soluble in water is more likely to be present in the interstitial fluids of soils. While phenol is biodegraded by soil micro-organisms under aerobic conditions this is not considered to take place anaerobic conditions.

Principal hazards

Phenol is highly toxic with a fatal dose of between 10–30 g although 1 g can be fatal in some cases (35). On this basis, a fatal dose for a child may be as low as 0.3 g.

Phenol poisoning under conditions of industrial exposure is associated with irritation of the mouth and throat, dark urine (due to the oxidation products of phenol) digestive disorders and Central Nervous System (CNS) effects. Severe chronic exposure may result in digestive disorders and CNS effects such as headaches, fainting and vertigo. The minimum daily dose for the onset of chronic symptoms appears to be about 1%. Phenols may also be carcinogenic.

Phenol is also readily absorbed through the skin on direct contact, causing white and blistered skin, or burning on prolonged contact. Loss of feeling may result in the affected area which may permit a more prolonged contact than would otherwise occur.

Phenol is toxic to fish and other aquatic life and may taint the flesh of fish at very low levels. The presence of phenol close to, or in, water supplies is considered undesirable for two reasons. Firstly, it can cause the formation of chlorinated phenols in chlorinated water supplies. These have an objectionable taste even at very low concentrations. Secondly, phenols present in soils in which plastic water pipes have been laid are able to attack the piping materials, having implications for both the structural integrity of water pipes as well as for water quality. Phenols may also affect concrete (36).

Some phytotoxic effects may occur in relation to high (> 1000 mg/kg) levels of phenol in soil while malodour at high concentrations may be a nuisance in some cases.

The principal hazards associated with phenols in respect of the redevelopment of contaminated land may therefore be summarized as the possibility of:

Ingestion of contaminated food and water
Problems associated with direct contact of phenol-contaminated soil
Potential water pollution problems
Material attack
Possible phytotoxic and malodour problems

Guideline levels

Planned uses	Trigger concentrations (mg/kg) air-dried soils
Domestic gardens and all uses with mains water services	5
All other uses	100

3.4.2 *Coal tars*

Sources of contamination

These are the main organic pollutants associated with coal carbonization. The chemical composition of coal tars is complex consisting of saturated and unsaturated aliphatic and alicyclic hydrocarbons, aromatic hydrocarbons (including polynuclear aromatic hydrocarbons), phenols, cresols, xylenols and related compounds, pyridine and other basic compounds, thiopen and related compounds and char (29).

Coal tar contamination of gasworks sites can be extremely variable with residual materials being present in some cases in underground pipes and tanks (tar wells) and extremely persistent deposits, for example, of tar and pitch also possible on some sites.

While extremely high concentrations of coal tars may be observed on former gasworks sites (Wilson and Stevens record a maximum concentration of 60% on the basis of studying a number of such sites (29)), the fractions responsible for particular hazards, for example, acute toxicity, skin contact hazards, account for a much smaller proportion of the total coal tar present (see below).

Principal hazards

Coal tars are a collection of different organic materials and the hazards associated with coal tars can be attributed to different fractions making up the whole.

Low molecular weight aromatics, such as benzene and toluene constitute about 2% of total coal tars. These materials are regarded as presenting an acute toxicity hazard since they are volatile substances with narcotic properties. They could present a short-term risk to investigators or worker on a site especially if present as residuals in underground pipes or vessels.

Other important constituents of coal tars are polynuclear aromatic hydrocarbons (PAHs) some of which are of concern with respect to their potential to cause cancer, and the presence of these substances in coal tars has led to an assessment of the potential hazards associated with both short and long-term contact with coal tars. Much of the evidence for skin irritation and long-term carcinogenesis of coal tars has come from studies on occupationally exposed individuals and animal experiments, making it difficult to draw exact parallels with relatively short duration contact with, such materials in contaminated soils. Nevertheless, the fact that coal tars contain PAHs of proven carcinogenic capacity to the extent of 5% of their total and the evidence of past industrial exposure has meant that visible evidence of coal tar contamination has been taken to indicate some degree of hazard of direct contact.

The inhalation of coal tar contaminated dust, particularly pitch, should also be avoided (although the link between coal tars and, for example, lung cancers was 'observed' only in workers occupationally exposed to fumes).

Ingestion hazards in relation to coal tars are not considered significant both because of the large quantities (> 50 g) of material which would need to be ingested to effect a fatal dose and also because chronic effects relating to the development of cancer are thought unlikely as PAHs are not considered to easily pass through the gut.

There is some concern that PAHs may be taken up by vegetables although evidence for this is conflicting and chronic effects via this route for PAH transfer are minimal for the same reason as for direct ingestion of coal tars.

However, coal tars do represent a problem in respect of water pollution since they can taint water supplies and the lower order aromatic fraction may chemically attack plastic piping. In addition, there is some evidence of phytotoxic effects on germination (at concentrations of 0.2%) and growth (at concentrations of 2%) while odour and discoloration of soils may be a nuisance if permitted to remain at high levels after redevelopment work.

The principal hazards associated with the contamination of sites with coal tars may therefore be summarized to include:

Direct skin contact with contaminated materials
Inhalation of contaminated dusts
Contamination of drinking water
Phytotoxic, odour and soil discoloration problems.

Guideline levels

Planned uses	Trigger concentrations (mg/kg) air-dried soil
Domestic gardens, amenity areas	200
Public open space	500
Industrial (no landscaping)	5000

These values do not take into account the hazard from skin contact with local tars.

3.5 Asbestos

Sources of contamination

Asbestos contamination of soils can arise in a number of ways including the contamination of sites using asbestos or asbestos-containing material in an industrial or constructional/insulation capacity or as a result of the disposal of waste materials containing asbestos.

A recent DoE guidance paper on asbestos-contaminated sites (37) suggests that the following sites may be contaminated with asbestos.

Former railway land (workshops, depots, sidings)
Heavy engineering sites (e.g. ship builders/breakers)
Old waste disposal sites (predating current legislation)
Sites used for manufacture of asbestos-bearing products
Scrap yards
Power stations

Principal hazards

The main hazard associated with asbestos is the inhalation of asbestos dust and fibres possibly leading to the development of respiratory disease. Asbestos related diseases include asbestosis, bronchial cancer and mesothelioma.

Asbestosis is a form of pneumonoconiosis and is a progressive lung fibrosis caused by the inhalation of asbestos dust. It is confined mainly to those industrially exposed over a long period. Bronchial cancer appears to be a complication of asbestosis having the same exposure characteristics. However, mesothelioma is a rare form of cancer of which asbestos is the most important single cause. The incidence of mesothelioma is not simply dose related and brief exposure to high concentrations of asbestos, particularly crocidolite, may carry a significant risk. Most mesotheliomas develop in individuals industrially exposed to asbestos although the occurrence of malignancies in the families of asbestos workers or individuals living close to asbestos factories underlines the potential hazard of the material in a wider environmental sense.

Four types of asbestos are normally considered hazardous in relation to asbestosis, namely crocidolite, amosite, chrysotite and anthophyllite. All four are hazardous, although the incidence of mesothelioma is linked to the type of asbestos involved in decreasing order of risk of crocidolite > amosite > chrysotile > anthophyllite.

In the context of contaminated land, the most significant hazard arises when fibres of dust are released into the atmosphere due to weathering, erosion or disturbance, (during de-commissioning, for example) or as a result of inadequate covering of materials disposed of onto landfill sites. Friable materials covered by granular material on a dry, well drained site represent a more hazardous situation than one in which asbestos fibres are confined in manufactured articles such as asbestos-cement boards, sheets or pipes or if a site is wet and the waste is covered by fine grained materials such as clay. Some difficulty may be experienced if asbestos-contaminated material is scattered over a wide area of a site and a closely spaced grid system of visual inspection may be necessary.

There is some controversy over the level of airborne asbestos which may constitute a hazard with respect to situations outside those of industrial exposure. Some authorities believe there to be no safe level of exposure, while others claim that a certain degree of exposure is inevitable in industrial societies and does not represent a more unacceptable health risk than many other readily accepted health hazards. Current DoE guidance (37) states that since it is not possible to quantify a hazardous airborne level, it is not possible to define that quantity of asbestos material present on a landfill site which gives rise to a hazardous level of asbestos in the air. Therefore the note is limited to providing advice on what reasonable measures can be taken to minimize risks. Such guidance includes the involvement of Environmental Health Departments and other specialist groups, attention to site security and possible methods of dealing with waste asbestos which has been disposed of in an unsatisfactory manner.

3.6 Combustible materials

Problems of combustion can arise in sites in which potentially combustible materials have been sorted, processed or dumped. Landfill sites, colliery spoil heaps and areas used for the storage of coal and coke in addition to sites in which soils have become contaminated with oil or other combustible liquids may be a cause for concern.

In many cases the potential for combustion is obvious because of a long history of, for example, tip fires. In other cases a lack of potentially combustible materials makes the hazard a negligible one. However, the difficulty arises in determining which of the sites intermediate in characteristics to these extremes represents a significant hazard.

Current DoE guidance (38) on these matters relate to combustible materials on a landfill site. It is suggested that a calorific value of 3000 BTU/lb (7.0 mg/kg) indicates that a potential hazard exists.

3.7 Gases

Gaseous emissions of toxic, flammable or asphyxiant nature, emanating from contaminated sites may represent both short and long-term hazards depending on the nature of the gas and of the development. The following is a brief summary of some of the important sources and characteristics of gases which may be encountered on contaminated sites. For a more detailed discussion on the problems associated with methane producing sites the reader is directed to Chapter 11.

Methane

Methane is a non-toxic gas which, in sufficient concentrations, is an asphyxiant. The gas can be generated, in association with other gases from the degradation of organic materials under anaerobic conditions. Methane is flammable at concentrations between 5–15% in air, and under some circumstances the accumulation of methane can lead to explosions. The presence of the gas in confined spaces is of particular concern therefore when accumulation takes place in or under buildings constructed on redevelopment sites since explosions can lead to loss of life and damage to property.

Carbon dioxide

Carbon dioxide is a toxic and asphyxiant gas, denser than air and therefore capable of accumulating in deep pits or excavations. It is produced in landfill sites by decomposition of organic matter or as a product of combustion. Carbon dioxide dissolved in water forms a highly corrosive liquid. The generally accepted limit for the concentration of carbon dioxide in inspired air is 0.5%.

Hydrogen sulphide

This is a highly toxic, flammable gas with a distinctive odour of rotten eggs. It has similar explosive properties to methane. Toxic effects are obtained at concentrations in excess of 400 ppm with subacute effects, irritation of the respiratory tract and eyes, observable at concentrations between 20–150 ppm. At concentrations greater than 700 ppm the presence of the gas is life threatening. The gas is also toxic to plants. Hydrogen sulphide is produced under anaerobic conditions, by microbial action on sulphate salts and is a well-known contaminant of kraft paper mill sites, oil refineries, coal carbonization sites and chemical works.

Carbon monoxide

Carbon monoxide is odourless, but highly toxic and flammable. It is produced during incomplete combustion of organic materials. The threshold limit value for the gas is set at 50 ppm although higher levels are tolerable over shorter exposure times than those relevant to TLV. The concentration range over which combustion is possible is 12–75%.

Sulphur dioxide

This is an irritating gas at concentrations between 6–12 ppm. It may be released during the combustion of contaminated materials (e.g. spent oxide) during the deliberate or accidental lighting of fires on former gasworks sites.

Petrol

Petrol vapour is a toxic (at concentrations of only 0.1%) substance which may occur as a result of spillage or arise from residual materials left on site. Petrol vapour is flammable within the range 1.4–7.6%.

Various volatile organic compounds

There is gathering evidence that volatile organic compounds, when present on former uncontrolled hazardous waste sites, can present risks to health not only to site investigators, but, if left untreated, to residents living close to sites and also potentially to occupiers or users of a site. Attempts have been made recently to document culprit material, which may include such substances as the lower aromatic compounds such as benzene and halogenated organic compounds such as dichloromethane, trichloroethane, methyl benzene and dimethyl benzenes (39). It appears that some compounds, which can exhibit irritant, narcotic or even carcinogenic properties are present in significant quantities in the atmosphere close to hazardous waste sites and attempts have been made to compare measured concentrations with threshold limit values and provisional limit in air values, the latter being a measure of the maximum constituent concentration considered safe in terms of continuous exposure in air in situations outside of occupational exposure.

3.8 Other contaminants

There is a vast number of materials which under certain conditions could be regarded as being significant contaminants of soils and groundwater, but for which no formal guidance, with respect to the redevelopment of contaminated land, exists as yet in the UK. Substances which might be considered include the polychlorinated biophenyls (PCBs) and dioxins, both of which are persistent and hazardous materials for which examples of soil contamination have already been documented and practical action has been demanded in respect of particular sites.

There are, in addition, a large number of organic chemicals used routinely in British manufacturing industry to produce paints, plastics, resins, synthetic fibres, pharmaceuticals, agrochemicals and pigments and which are constituents of waste streams destined for discharge to sewer or ultimate disposal to landfill. Some European countries have embarked on developing 'action' guidelines to deal with contamination problems associated with 'organics' while in the UK there have been modest moves to proceed along similar lines.

Prospective changes in industrial structure may mean that substances such as acrylonitrile and styrene rather than cyanide and cadmium become the common soil contaminants of the future, and despite recent suggestions that appropriate guideline figures for some contaminants may never be derived (40) the implications of current and future soil contamination must mean that the development of guidelines relevant to particular types, or specific contaminants, should remain an important area of endeavour.

References

1. Peterson, P.J., Benson, L.M. and Zieve, R. (1981) Metalloids. In: *Effect of heavy metal pollution on plants*, Vol. 1, (ed. N.W. Lepp) Applied Science, London, 279–342.
2. Thornton, I. (1981) Geochemical aspects of the distribution and forms of heavy metals in soils. In: *Effect of heavy metal pollution on plants*, Vol. 2, (ed. N.W. Lepp) Applied Science, London, 1–33.
3. Sinclair, W.A., Stone, E.L. and Scheer, C.F. (1975) Toxicity to hemlocks grown in arsenic contaminated soil previously used for potato production. *Hort. Sci.*, **10**, 35–36.
4. Aggett, J. and Aspell, A.C. (1980) Arsenic contamination in an apple orchard. *Environ. Pollut.*, **22**, 39–46.
5. Porter, E.K. and Peterson, P.J. (1977) Arsenic tolerance in grasses growing in mine waste. *Environ. Pollut.*, **14**, 255–267.
6. Department of the Environment, (1980) Cadmium in the environment and its significance to man, Pollution Paper No. 17, HMSO, London.
7. Webber, J. (1981) Trace metals in agriculture. In: *Effect of heavy metal pollution on plants*, Vol. 2, (ed. N.W. Lepp), Applied Science, London, 159–184.
8. Heeps, K.D. (1982) The reclamation of a disused sewage works, *Public Health Eng.* **10**, 213–218.
9. Department of the Environment, (1984) Cadmium bearing wastes, Waste Management Paper No. 24, HMSO, London.
10. Barltrop, D. (1986) Evaluation of cadmium exposure from contaminated soil. In: *Proc. Conf., Contaminated Soil*, Utrecht, The Netherlands, Martinus Nijhoff, Dordrecht, 169–180.
11. Alloway, B.J. and Morgan, H. (1986) The behaviour and availability of cadmium, nickel and lead in polluted soil. *Ibid*, 101–113.
12. Thornton, I. (1986) Metal contamination of soils in UK urban gardens: Implications to health. *Ibid*, 203–209.
13. Cooney, P. (1980) Heavy metal contamination: Background levels and site case histories in the London Borough of Greenwich. Papers to a workshop, Case histories of contaminated land—local authority viewpoints, Research Note 16, JURUE, Birmingham.
14. Piscator, M. (1979) Metabolism and effects of cadmium. In: *Proc. Conf. Management and control of heavy metals in the environment*, London, CEP Consultants Ltd., Edinburgh, 1–7.
15. Simms, D.L. and Beckett, M.J. (1985) Contaminated land: Risk analysis and safety margins. Paper presented to Conf., Contaminated Soil, Utrecht, The Netherlands.
16. John, M.K. and Van Learhoven, C.J. (1976) Differential effects of cadmium on lettuce varieties. *Environ. Pollut.*, **10**, 163–173.
17. Davies, B.E. and White, H.M. (1980) Cadmium and lead contamination of soils and vegetables in relation to historic base metal mining. In: *Proc. Conf., Reclamation of Contaminated Land*, Eastbourne, SCI, London, C8/1–C8/8.
18. Lindsay, D.G. (1980) Toxic elements in food. Ibid, C1/1–C1/7.
19. Page, A.L., Bingham, F.T. and Chang, A.C. (1981) Cadmium. In: *Effect of Heavy Metal Pollution on Plants*, Vol. 1, (ed., N.W. Lepp) Applied Science, London, 77–109.
20. Roberts, T.M. *et al.* (1974) Lead contamination around two secondary lead smelters in downtown Toronto: Estimation of ongoing pollution and accumulation by humans. Institute for Environmental Studies, University of Toronto, Pub. No. EE-1.
21. Harris, M.R. (1981) Concentration of lead in the atmosphere and soil measured in the vicinity of a secondary lead smelter. *Environ. Tech. Lett.*, **2**, 233–242.
22. McCarthy, M.J. (1980) Reclamation of a refuse tip for open space and housing development. In: *Proc. Conf., Reclamation of Contaminated Land*, Eastbourne, SCI, London, B8/1–B8/11.
23. Koeppe, D.E. (1981) Lead: Understanding the minimal toxicity of lead in plants. In: *Effect of Heavy Metal Pollution on Plants*, Vol. 1, (ed. N.W. Lepp) Applied Science, London, 55–76.
24. Page, R.A. and Swires-Hennessy, E. (1986) A safe level for lead in soil and dust. In: *Proc. Conf., Contaminated Soil*, Utrecht, The Netherlands, Martinus Nijhoff, Dordrecht, 233–246.
25. Dickinson, N.M. and Lepp, N.W. (1984) Pollution of tropical plantation crops by copper fungicides: A copper budget for a Kenyan coffee plantation. In: *Proc. Conf., Environmental Contamination*, London, CEP Consultants Ltd., Edinburgh, 341–346.
26. Heeps, K.D. and Pike, E.R. (1980) Reclamation of a disused sewage farm. In: *Proc. Conf., Reclamation of Contaminated Land*, Eastbourne, SCI, London, B6/1–B6/10.

27. Freedman, B. and Hutchinson, T.C. (1981) Sources of metal and elemental contamination of terrestrial environments. In: *Effect of Heavy Metal Pollution of Plants*, Vol. 2, (ed., N.W. Lepp) Applied Science, London, 35–94.
28. Collins, J.C. (1981) Zinc. In: *Effect of Heavy Metal Pollution* on Plants, Vol. 1., (ed., N.W. Lepp) Applied Science, London, 145–169.
29. Wilson, D.C. and Stevens, C. (1981) Problems arising from the redevelopment of gas works and similar sites. AERE Report R10366, HMSO, London.
30. Department of the Environment (1976) Heat-treatment cyanide wastes. Waste Management Paper No. 8, HMSO, London.
31. Rice, R.G. (1984) Purification and recycling of groundwater contaminated with petroleum products and cyanides—the Karlsruhe (FRG) drinking water treatment plant. In: *Proc. Conf., Management of Uncontrolled Hazardous Waste Sites*, Washington DC., HMCRI, Silver Spring, Maryland, 600–603.
32. Roberts, T.M. and Gemmell, R.P. (1980) Establishment of vegetation on gas works waste. In: *Proc. Conf., Reclamation of Contaminated Land*, Eastbourne, SCI, London, F11/1–F11/10.
33. Personal Communication (1985) London Borough of Greenwich.
34. ECOTEC Research and Consulting Ltd., (1985) The re-use of landfill sites. A report to the EEC Commission of the European Communities, ECOTEC Research and Consulting Ltd., Birmingham.
35. Patty, F.A. (1963) *Industrial Hygiene and Toxicology*. Vol. II., Interscience, New York.
36. Smith, M.A. (1986) Effect of phenol on concrete, *Mag. Concrete Research*. (in press).
37. Interdepartmental Committee on the Redevelopment of Contaminated Land, ICRCL (1985) Asbestos on contaminated sites. ICRCL Paper 64/85, 1st edition, DoE, CDEP, London.
38. ICRCL. (1984) Notes on the fire hazards of contaminated land. ICRCL Paper 61/84, 1st edition, DoE, CDEP London.
39. James, S.C., Kinman, R.N. and Nutini, D.C. (1985) Toxic and flammable gases. In: *Contaminated Land*, (ed., M.A. Smith) Plenum, New York, 207–255.
40. Beckett, M.J. and Simms, D.L. (1986) Assessing contaminated land: UK Policy and practice. In: *Proc. Conf., Contaminated Soil*, Utrecht, The Netherlands, Martinus Nijhoff, Dordrecht, 285–293.

4 Appropriate site investigations

D.W. LORD

4.1 Introduction

The logical sequence of site investigation to incorporate a blend of scientific necessity and economic viability follows five distinct stages:

(i) Recognition that a investigation is needed and establishment of the basis for that investigation
(ii) Design of a site sampling/analytical programme
(iii) On site sampling, observation and testing
(iv) Analysis of materials taken from the site
(v) Interpretation of analytical data.

4.2 Recognition that an investigation is needed and establishment of the basis for that investigation

4.2.1 *Recognition*

The creation of the right conditions for investigation is fundamental to achieving the optimum design of site investigation. A major factor in creating those conditions is early recognition that investigation is needed.

If the first sign that a site is contaminated is the discovery of some highly visible pollutant e.g. tar, during construction of foundations for houses then proper investigation is heavily constrained by the economic factors surrounding the housebuilding. In those circumstances investigation cannot be designed to provide information for a wide range of possible options in site reclamation. The constrained investigation can only be related to the development in progress and indeed the speed at which the development must proceed to be economic.

There is therefore pressure on the constrained investigation to offer not only a safe solution to the problem but one which can be carried out very rapidly.

This latter requirement could lead to high remedial costs and indeed may even point to the fact that a different form of development could have been more appropriate for this site.

The Department of the Environment's Guidance Note (1) on assessment and redevelopment of contaminated land stresses: 'The aim therefore should always be to check whether a site is contaminated before deciding on the form of development'.

An all-embracing approach to identifying contaminated sites and providing information on potential contamination has been described as a 'prospective survey' (2, 3).

4.2.2 *Prospective Surveys*

JURUE (2) defines the prospective survey as 'a desk exercise which aims to establish the former uses of an area or a site and from this information to assess the probability of finding contamination on specific sites in that area'.

Such surveys can provide information at two levels. In the first instance they can enable sites to be recognized as potentially contaminated and, following more intensive study, they can identify the possible contaminants and the parts of a site where contamination might be most and least expected.

In the study of a large area, say a local authority district, relevant information can be used by the experienced investigator to classify sites on their likelihood of being contaminated. JURUE (2) describe eight classifications ranging from 'High probability of being contaminated' to 'Unknown'.

With such a classification on record, when redevelopment is being considered appropriate responses can be triggered.

The information required in area prospective surveys has been reviewed by JURUE (2) and Barry (3) and one approach to construction of a large area contaminated land register has been described by Duffy (4).

4.2.3 *Site specific prospective surveys*

A site specific prospective survey would consider a wide range of current and historical information to evaluate:

(a) The activities carried out on the site
(b) The location of those activities
(c) The possible contaminants released by those activities.

Sources of information for 'desk study' have been reviewed by Dumbleton (5, 6) and include:

Maps—ordnance survey, town, geological etc.
Photographs—air and ground
Local literature—local history, local newspapers etc.
Directories—trade, street etc.
Technical literature
Local, regional and central government bodies—local authorities, water authorities, HSE etc.
Local knowledge—residents in the area etc.

A Draft British Standard Code of Practice (7) has considered in detail the value and applicability of this range of sources of information to contaminated land investigation, and has concluded that, whilst a high degree of value judgement is necessary in making most efficient use of information sources, and whilst studies of literature, maps etc., could, if uncontrolled, amount to too much unproductive work, site specific prospective surveys are potentially very valuable.

Many site specific prospective surveys can be viewed as attempting to answer four questions:

(i) What significant industries have operated on the site?
(ii) Are any of the processes used on site, producers of contaminants?
(iii) For each 'contaminating' process or part process what were:
 (a) The positions on site?
 (b) The raw materials that would be regarded as significant contaminants and where and how were they stored?
 (c) The end products that would be regarded as significant contaminants and where and how were they stored?
 (d) The position of other storage vessels?
 (e) The locations of any leakages that may have occurred in pipelines or storage vessels?
 (f) The waste materials which would be regarded as significant contaminants and where and how were they stored?
(iv) Finally, can we, from the answers to the previous questions, produce:
 (a) A plan indicating potential contaminant 'hot-spots'?
 (b) A list of potential contaminants?

In answering question(i) it can be useful to have in mind a list of industries which potentially fall into the 'significant' category, such a list would include:

Table 4.1

Coal carbonization works
Alloy smelting and casting works
Iron and steelmaking plants
Tar and refineries (Figure 4.1)
Paint works
Scrapyards (Figure 4.2)
Tanneries
Petroleum refineries and storage facilities
Sewage works, farms
Foundaries
Printing works
Bleach and dyeworks
Landfill sites—domestic, commercial, industrial
Manufacturing chemical plants
Plating works
Metal mining areas
Munitions works
Solvent recovery plants
Dockland storage areas

Figure 4.1 A derelict tar works

In assessing whether the listed industries or other industries incorporate processes which may produce contaminants consultation with those who have worked in similar industry coupled with the study of technical literature should enable the experienced investigator to formulate possibilities.

The greatest difficulty lies in attempting to answer question (iii). This question implies a demand for a detailed knowledge of process layout and of change in layout over the life of the site. In instances where the industry concerned still exists elsewhere, detailed plans and descriptions may be available.

In the author's experience such detail is likely to be incomplete, but often sufficient detail is available to indicate certain process locations, and hence certain potential hot-spots.

One example of the level of information which can be uncovered to enable site investigation to be designed is now described.

A methyl methacrylate recovery plant had been established in a disused cotton mill (where textiles had previously undergone physical processing but no chemical treatment). Scrap polymethyl methacrylate (e.g. production off-cuts) was brought onto the site and stored in one area above uncovered ground. The scrap was milled to produce a granular material and the granular polymer was thermally cracked in pot stills heated by a lead-based heat sink.

The cracking process was designed to produce the monomer methyl methacrylate but other by-products—including aliphatic esters and acids would be produced dependent on conditions. (Analysis of waste from the stills during the life of the site had shown the waste to be predominantly methyl methacrylate.)

The methyl methacrylate fraction from cracking was cleaned by washing with sulphuric acid and neutralized with sodium hydroxide.

The cleaned monomer was fractionated for final purification.

At the time of prospective survey the plant buildings were partly intact and investigation (e.g. site inspection and discussion with former staff) revealed the exact sites of pot stills, fractionating column etc., and that the following potential contamination routes existed:

(i) Storage of raw material on open ground
(ii) Leakage from pot stills through poorly maintained concrete flooring
(iii) Spillage during transport of recovered monomer in open drums from pot stills to the fractionating column, some distance away

Figure 4.2 A scrap yard. © Fox Photos Ltd.

(iv) Leakages from acid and alkali washing processes
(v) Leakages from wastes stored in drums on open ground awaiting off site disposal (e.g. acid wash liquor, alkali wash liquor, still residues)
(vi) Dumping of waste materials (including lead from stills) on two parts of site
(vii) Leakages from drummed product stored on open ground awaiting off site transport.

Thus potential hot spots were identified. Furthermore the knowledge of the process allowed the following list of primary potential contaminants to be constructed:

Polymethyl methacrylate
Methyl methacrylate
Lead
Sulphuric acid
Sodium hydroxide

Overall there was adequate information to design a programme of site sampling/analysis.

Similarly there are other published examples of sites where processes can be linked to contaminants. Aldred and Lord (8) described a cotton dye/bleach works where waste from the dye process was separated and stored/disposed in three areas of the site. Likely hot-spots were identified and the following list of potential contaminants prepared:

Sulphate
Phenols
Dyestuffs
Copper
Chromium
Arsenic
Lead
Cadmium
Zinc

The relationship between processing and site contamination in coal gas production is documented by Wilson and Stevens (9) and a number of examples of site study are given.

Pearce and Finnecy (10) have described the processes in steel production likely to give rise to ground contamination and have identified potential site contaminants.

In a series of publications the Department of the Environment (11–13) have described how and with what landfill sites, scrap yards and sewage works may be contaminated.

It is worth noting that the latter three types of site can be very different in extent of contamination from the process contaminated sites referred to earlier.

For example in an old landfill, prospective surveying is unlikely to reveal areas in which particular wastes have been tipped—although with the advent of requirements of The Control of Pollution (Special Waste) Regulations (14) this may be possible in the future.

Thus although a list of potential contaminants can be constructed, potential hot-spots are unlikely to be postulated from prospective surveying.

4.2.4 *Site reconnaissance*

The site-specific prospective survey can be a study of very varying depth and resource requirement.

It may at one extreme be a simple discovery that the site in question had been for example an operational dye works and hence may be contaminated and thus should be subject to further investigation before redevelopment is planned in detail, i.e. the survey simply recognizes that investigation is needed. At the other extreme the prospective survey can be intended to provide the optimum level of detailed information for design of a site sampling/analytical programme.

The latter form of prospective survey will certainly have left the area of 'desk study' to embrace site reconnaissance (i.e. exploration of the site surface) as an essential feature of identifying potential contaminants and contaminated areas.

Figure 4.3 Site reconnaissance

Site reconnaissance is needed to observe:

(i) The present state of the site, its topography and the influence of demolition works etc., in spreading contamination
(ii) Clues indicating the location of previous site buildings
(iii) The general nature of visible site fill
(iv) The presence of surface waste deposits
(v) The presence of odours
(vi) The presence/absence of vegetation
(vii) The relationship to nearby sites, watercourses etc.

4.3 Design of site sampling/analytical programme

4.3.1 *The hazard associated with contaminants*

The scope of the site sampling/analytical programme and ultimately the extent of remedial works needed will, of course, vary greatly from site to site.

An appreciation of the link between the risk presented by hazardous contaminants to site development and the nature of the development itself is needed to design the investigation which will be appropriate for that site.

Identification of the hazardous contaminants which may be present in the site will have been the subject of the prospective survey. Indeed their definition as potentially hazardous is determined by the likelihood that they present a risk to:

(i) Site workers, e.g. by skin contact or by inhalation
(ii) Site work, e.g. by explosion or fire as a result of 'spark' generated during excavation
(iii) Building foundations, e.g. attack on concrete
(iv) Building/site services, e.g. attack on water mains, cables etc.
(v) Buildings, e.g. combustion of contaminant material
(vi) Building occupants, e.g. gas build-up in confined spaces
(vii) Garden, allotment users e.g. by skin contact, and the consumption of contaminated vegetables
(viii) Children, e.g. skin contact, ingestion of fill
(ix) Neighbouring sites, e.g. movement of leachate
(x) Plant growth.

Clearly many different combinations of those risks may be present in any one site.

4.3.2 *'Sensitivity' of site use*

Attempts have been made to grade sites on scales of 'sensitivity' according to their intended use. Such that those defined as 'most sensitive' would require

Table 4.2

Greatest sensitivity	Residential development with gardens
	Development for agriculture or allotments
	Amenity, recreational parkland use
	Industrial development—offices, factories
Least sensitivity	Car parks

greatest investigation and those defined as 'least sensitive' would require least investigation.

One such scale of sensitivity related to environmental health and well being that has been proposed (7) is given in Table 4.2.

In many cases, such a scale is likely to be a reasonable guide to the depth of investigation needed. However each potential hazard must be considered individually to properly evaluate the likely risk. Consider for example the following simplified situations. Site fill contaminated by 10 000 mg/kg of lead (expressed as Pb) occurring as a water insoluble salt would properly not be considered as desirable for use as garden soil. However the same fill could properly be covered with hardcore and asphalt for car park construction.

It is worth noting that even where such 'safe' options as the latter are adopted, there is also a necessity to ensure that details of site contamination are stored for access in case a future redevelopment may occur.

In the former case (i.e. that of potential garden development), once the likelihood of lead contamination was indicated by prospective survey extensive site sampling/analysis would be needed to identify the lateral and vertical spread of lead contamination. In the case of the car park development, only a broad indication of the lead contamination would be needed, since the redevelopment would be deemed appropriate if the lead in fill level was say, up to 1000 mg/kg or up to 20 000 mg/kg. Conversely if the fill were contaminated with, say, 3000 mg/kg of petroleum spirit and hence likely to exhibit a 'flash point', then it would be no more desirable for housing with gardens than for car park construction, where in both cases vapour ignition could occur during reclamation, and vapour penetration occur after development. Thus once the possibility of petroleum spirit contamination was revealed, a similar extent of site sampling/analysis would be needed irrespective of intended use.

Beckett (15) attempted to categorize the hazard associated with various contaminants according to site use and produced Table 4.3.

4.3.3 *Multi-disciplinary approach*

It is precisely because of the links between past use of site, present use of site, hazardous contaminants, risks to development, intended development and of course cost, that a multi-disciplinary approach to site investigation is needed.

The range and complexity of technical and other issues is such that no single scientific discipline or profession can deal with them.

Table 4.3

Hazard	*Typical end uses where hazard may exist*	*Priority contaminants*
Direct ingestion of contaminated soil by children	Domestic gardens, recreational and amenity areas	Arsenic Cadmium Lead Free cyanide Coal tars Phenols Sulphate
Up take of contaminants by crop plants	Domestic gardens, allotments and agricultural land	Cadmium Lead pH
Phytotoxicity	Any uses where plants are to be grown	Sulphate pH Copper (total & available) Nickel (total & available) Zinc (total & available)
Attack on building materials and services	Housing developments, commercial and industrial buildings	pH Sulphate Sulphide Chloride Coal tars Phenols Mineral oils
Fire and explosion	Any uses involving the construction of buildings	Methane Sulphur Potentially combustible Materials, (e.g. coal dust, Oil, tar, pitch, rubber)
Contact with contaminants during demolition, clearance and construction work	Hazard mainly short-term (to site workers and investigation teams)	Coal tars Phenols Oily and tarry substances Asbestos Radioactive materials
Contamination of groundwater	Any uses where possible pollution of groundwater may occur	Phenols Cyanide Sulphate pH Soluble metals

Notes from (15)

(i) These hazards are not mutually exclusive and combinations of several hazards may need consideration.

(ii) Other contaminants may need to be included where the site history has identified former uses likely to have introduced them.

The multi-disciplinary team established to effect the planning and execution of site investigation will depend upon the circumstances of the particular site and the interested organizations.

However, for example, in the author's experience, many local authority staffed site investigations centred around a team of an engineer, an Environmental Health Officer, a planner and Public Analyst, and drawing as needed on the other expertise within and outside the local authorities; have proved very effective.

4.3.4 *The aim of sampling/analysis*

The aims in sampling and analyzing site materials are to:
(i) Identify the highest levels of contamination and (ii) describe the overall spread of contamination both laterally and vertically.

The ideal sampling plan would therefore have sufficient sampling points which were distributed such that firstly, hot-spots were located and secondly, the overall extent of contaminated area could be indicated. Given that a site may have a large number of individual contaminants generated from a number of different processes in different locations, and have been subject to several mechanisms of spread of contamination e.g. groundwater movement, effects of demolition etc., it is not surprising that a mathematical model describing an appropriate contaminated site sampling plan does not exist.

However the following must be defined at the design stage:

(i) The sampling pattern or patterns that should be used
(ii) The number of samples that should be taken
(iii) The depths at which the samples should be taken
(iv) The number of stages (i.e. sampling time intervals) there should be.

4.3.5 *Sampling patterns and sample numbers*

4.3.5.1 *Composite sampling*. Widely used in agricultural/horticultural land investigation are the 'W' and 'X' patterns of sampling. The general premise adopted is that the distribution of plant nutrients is relatively homogeneous. Along the outline of a figure 'W' or 'X' a number of spot samples are taken, then bulked and mixed to provide one sample for analysis. The distribution of sampling points is likely to be inadequate to provide location of hot-spots, and in any event 'high' contaminant levels will be lost in mixing of the spot samples. Thus in the majority of contaminated land investigations these well established patterns are unlikely to be useful. They may have some application in certain areas e.g., ex-domestic refuse landfill site but their ease of use is generally outweighed by the disadvantage of obscuring 'high' levels of contamination by bulking.

As a general rule any sampling plan involving the use of composite rather than 'spot' samples will find limited use in contaminated land investigation.

4.3.5.2 *Random Sampling*. Since one objective of the sampling strategy is to locate hot-spots, then a sampling plan with a definable level statistical confidence in identifying such locations would appear to be appropriate to meet that objective. Random sampling techniques offer a vehicle for estimation and prescription of sampling error or the probability of locating hot-spots using binomial/hypergeometric sampling theory (16).

In simple random sampling, sampling points are randomly located over the whole area of the site. In such a design the site is usually divided on paper into a fine grid whose co-ordinates are allocated numbers. Random number tables are then used to select sampling points. Random sampling as in the sense of numbered lots selected from batches from a production line is not of course feasible in a site investigation. Nevertheless it is arguable that this approximation to random sampling justifies the consideration of the statistical confidence limits of the chosen plan.

Bell *et al.* (17) give examples from binomial distribution tables of the probabilities of locating hot spots at different sampling frequencies in Table 4.4.

This table of probabilities must be viewed in a two-dimensional sense, such that for example if 1% of the total site area is contaminated, then the sinking of 50 trial pits will enable a 39% chance of locating that contamination. Whereas if the site has 10% of its total area contaminated a 50 trial pit strategy will have a 99% chance of locating that contamination.

However in practice simple random sampling patterns tend to produce clumping of sampling points and significant areas of the site may therefore remain not sampled.

Clearly to have a high level of confidence in the sampling pattern, a relatively large number of sampling points is needed, equally clearly the larger the number of sampling points, the greater the cost of sampling and analysis. It is quite proper that site investigation should be conducted to a high level of confidence, but statistical considerations need to be married to experience in investigation and to knowledge of the site. The major weaknesses of the simple random sampling system are that, firstly it ignores the information gained from prospective survey concerning location of hot-spots, and secondly it allows no alteration of strategy based on experience or observation.

One other weakness in simple random sampling, that of clumping of

Table 4.4 Probability of locating contaminated areas of various sizes by random sampling

Contaminated areas as % of total site area	*Sample numbers* 10	30	50
1	0.10	0.26	0.39
5	0.40	0.79	0.92
10	0.65	0.96	0.99
25	0.94	1.00	1.00

sampling points, can be overcome using the modification of stratified random sampling. The site is divided into strata or cells and a number of samples chosen at random within the strata. The nature of the scheme still allows a valid estimate of sampling error to be made and more even site coverage is obtained.

Stratified random sampling is close in its concept to the most commonly used sampling pattern to achieve even coverage of the site—that of regular grid sampling, which is considered in the next section.

4.3.5.3 *Grid sampling*. This pattern is commonly used in determining the engineering suitability of the ground (18), offering an easy means of mapping and locating sampling points once the dimensions of the square grid are decided upon.

Wilson and Stevens (9) state that:

> Sampling grids used on gas works sites by different organizations have generally utilized spacings of between 20 m and 50 m between adjacent positions, with a 25 m grid being fairly common. For small sites a closer grid spacing may be required to achieve a reasonable minimum number of sampling points, say in the range 10–20 m.

Kelly (19) advises that for relatively clean sites (i.e. other than virgin land) a 100 m grid is an acceptable compromise between cost and information requirements. For more heavily contaminated sites (or sites on which 100 m grid has revealed the need for more testing) a 50 m grid is appropriate. Closer grids would be appropriate for small sites. Waterhouse (20) advocates the following grid sizes related to site size.

Table 4.5

Site size (hectares)	*Grid size m*
0.5	10
5	20
16	30

Lowe (21), in describing the investigation of 400 hectares of land at Thamesmead indicates grid sizes in various investigations ranging from 10 m to 100 m.

Carpenter *et al.* (22) recommended grid sampling and as a rough guide refer to:

> a site of
> 0.5 hectares would require a minimum of 5 sample locations
> 1.0 hectares would require a minimum of 9 sample locations
> 5.0 hectares would require a minimum of 20 sample locations

They also state that

> Unless there are strong reasons for not doing so the trial pits should be dug to a depth of 3 metres and 'spot' samples taken at 0.15, 0.5, 1.0, 2.0 and 3.0 m.

Table 4.6 Grid size (m) to facilitate location with 95% confidence of contamination of 5% and 10% of site area

Area of Site	*5% of area*	*10% of area*
0.5	9	13
1.0	13	18
1.5	16	22
2.0	18	26
2.5	20	29
3.0	23	31
4.0	26	36
5.0	29	41
6.0	32	45
7.0	34	48
8.0	37	52
9.0	39	55
10.0	41	58
11.0	43	61
12.0	45	64

Smith and Ellis (23) carried out an investigation on a gasworks site varying grid size from 75 m down to 6.25 m. From a study of the analytical data on samples taken at those grid spacings they concluded that a sampling grid of 25 m was adequate for an overall appraisal of the degree of contamination.

One approach towards choice of grid size is described by Bell (17) and noted in the draft BS Code of Practice (7).

The statistical basis of simple random sampling is related by Bell to regular grid sampling. The assumption is made that the random sampling density at the higher confidence levels dispersed over a regular grid would neither increase nor decrease the probability of locating contamination. The probabilities shown in Table 4.4 are thus assumed to apply to regular grid sampling. Thus for example if it is required to have 95% confidence in locating an area of contamination on a site then the grid sizes given in Table 4.6 would be appropriate.

This table indicates that for a 5 hectare site with an area of contamination of 5% of the total area, a grid size of around 29 m would be needed to give 95% confidence of locating the contaminated area.

However, this method allocation of a simple regular grid to a site ignores the experience of the site investigators and the details of processes and their locations obtained from historical survey.

The process of designing the sampling grid must marry together the knowledge of the site with the implications of the confidence levels described above.

The above approach to grid sampling is one way of giving some perspective to the statistical confidence one might have in a grid sampling plan. An approach to estimating not grid size, but the number of samples needed is now given.

4.3.5.4 *General approach to estimating sample numbers.* For environmental sampling in general, the American Chemical Society's Committee (ACS) on Environmental Improvement (24) has recommended an approach which is essentially the same as that for estimating numbers of water samples needed (in water quality investigations) by the DoE/NWC Standing Committee of Analysts (25).

The number of samples to be taken is related to the standard deviation of the sample population and a prescribed level of error as follows:

$$N = \left(\frac{t \cdot \sigma_p}{e}\right)^2$$

where

N = number of samples
σ_p = the estimate of the standard deviation of sample population
e = the tolerable error in the estimate of the mean value of the parameter of interest
t = value of Student's 't' distribution

The standard deviation of the population σ_p can be calculated from

$$\sigma_o^2 = \sigma_m^2 + \sigma_p^2$$

where
σ_o = overall standard deviation
σ_m = standard deviation of the measurement (i.e. that associated with sub-sampling and analysis)

In most cases σ_m is likely to be small compared to σ_p, so that an estimate of overall standard deviation σ_o will suffice as an indication of the value of the standard deviation of the sample population σ_p.

However by careful design of sampling and laboratory work, σ_m can be independently estimated, if required, to give a more precise estimate of σ_p.

The ACS recommended that prior experimentation to provide tentative estimates of σ_o and σ_m should involve not less than seven measurements on similar samples and duplicate samples respectively.

This general approach to estimating samples numbers has been expanded upon in a report published by the US Environmental Protection Agency (26).

The report recommends that the approach be used as the procedure 'for determining the number of samples required to meet a predetermined precision ... and the number of sample points required in the systematic sample design'.

For ease of use the EPA report converts the equation to the following form:

$$N = \left(\frac{\mathrm{CV} \cdot t}{p}\right)^2$$

where

N = number of samples required
CV = coefficient of variation
t = appropriate Student's t value which will generally be close to 2
p = allowable margin of error expressed as a %

To use the equation to indicate the estimate of sample numbers one needs

(a) An estimate of the coefficient of variation (CV)
(b) An agreed margin of error.

Mason (26) quotes a number of studies showing CV's in the range of 1% to 850%. Although for purposes of estimation, a CV of 65% is recommended for general use i.e. where no prior knowledge of CV exists.

Barth and Mason (27) indicate a small number of studies that have been so designed that the total number of samples taken can be identified to be within one or more populations and hence allow an estimate of CV for each population.

However in the author's experience, most studies are not so designed, and in the case of site studies carried out by this laboratory, and by others, the coefficient of variation (i.e. calculated from the overall standard deviation) for individual contaminants for the site as a whole can commonly be found in the range 50 to 250% or more.

If no prior knowledge of contaminant variability is to hand it would seem not unreasonable in the author's view to use a CV of 100 to 150% to assist in prediction of sample numbers needed.

The effect on sample number estimates of differing choice of the coefficient of variation within error bands of ∓20% to ∓30% is shown in Table 4.7.

Thus the table shows, for example, that for data on a particular contaminant exhibiting a coefficient of variation of 100% to achieve an estimate of mean contaminant concentration within ∓20%, 100 samples would be needed.

The statistical approaches to sampling outlined in 4.3.5.2, and 4.3.5.4 can be regarded to some extent as complementary.

Table 4.7

Coefficient of variation %	*Number of samples required to control error within limit specified*		
	∓ 20%	∓ 25%	∓ 30%
75	56	36	25
100	100	64	44
150	225	144	100
200	400	256	178
300	900	576	400

The first approach (4.3.5.2) is concerned with determining the confidence that can be achieved in locating contamination.

The second approach is concerned with determining the confidence that can be achieved in defining the numerical value of that contamination.

By taking cognizance of both approaches an estimate of:

(a) the number of trial pits needed, and
(b) the number of samples needed can be made.

It is stressed though that these two approaches offer only mathematical assistance, in defining sampling investigation design. They can provide valuable assistance but it would be unwise to rely entirely upon such numerical approaches at the expense of overall knowledge of the site and of investigation methods – see section 4.3.7.

4.3.6 *Sampling depth*

Contaminants deposited on the site surface during its life may migrate to depth, and indeed contaminants may have been released at depth through underground storage vessels etc. The depth at which knowledge of the contaminants is required can depend upon the site development intended, for example where site services are planned at 1.5 m below site surface; or on other factors, e.g. migration to adjacent sites.

Knowledge of site development options can assist greatly in defining the sampling depth needed. On some sites, an undisturbed clay layer may lie relatively close to the surface. Since this presents a general barrier to contaminant movement, it will in most of those cases prescribe the maximum sampling depth. As a general rule for many types of common developement e.g. housing, schools, recreational, a maximum sampling depth of 2.5 m or to undisturbed clay would be appropriate. Samples of fill should be taken at surface and at several depths and/or in reaction to observation during sampling. Fixed interval sampling e.g. 1m intervals can usefully be related to remedial works—say removal of 1 m surface layer, but should not be pursued at the expense of sampling unusual strata observed during excavation.

In general it is likely that a minimum of three samples per sampling point, i.e. per sampling pit, will be needed to provide adequate information on vertical spread of contamination.

4.3.7 *Sampling quality assurance*

4.3.7.1 Sampling techniques are discussed in section 4.4. In the evaluation of site contamination the sample plays a unique role. Individually samples must present an unbiased picture of a point on the site and collectively must represent the site as a whole. Some practical quality assurance procedures to ensure the integrity of the sample are:

(a) the use of clean and uncontaminating sample containers
(b) the use of clean and uncontaminating sample collection devices
(c) effective cleaning of sample collection devices between samplings
(d) effective storage of samples prior to analysis.

These procedures are expanded upon in sections 4.4 and 4.5.

4.3.7.2 Later in this chapter Analytical Quality Assurance (AQA) is discussed. AQA is concerned with the demonstration and maintenance of the quality of data produced in the individual laboratory by sub-sampling and analytical techniques. However the major source of uncertainty in contaminant analytical results is likely to result from the sampling of the material rather than from its sub-sampling and analysis. This is well demonstrated in other fields (28, 29) and explains the large coefficients of variation referred to earlier in section 4.3.5.4.

To achieve a degree of certainty in overall measurements which have a large coefficient of variation, relatively large numbers of samples are needed and this has already been indicated.

By careful design of sampling, sub-sampling and analytical work the components of variance attributed to each source of error can be measured. One investigation referred to later in section 4.5.3 was so designed. In most cases, though, such detailed experimental design will not be needed. There is, however, a case for the measurement of the variance associated with duplicate sample analysis. Such measurements can assist:

(a) in identifying whether the whole of the sample data can be treated as one population or several populations
(b) in assessing the uncertainty of individual measurements (see section 4.6.4.), as well as
(c) in comparing sample variability with analytical/sub-sampling variability and identifying any need for improvement in the latter.

Clearly the usefulness of duplicate sample analysis is dependent on the numbers of such duplicates available and in general a minimum of 7 replicated from any one population will be needed to produce a reasonable estimate of the variance. To be useful in the context of the site as whole in particular in relation to (a) above and to be useful as a general quality assurance mechanism it is suggested that a minimum of 5% of all samples taken should be duplicated, i.e. a second sample taken at the same point. This, to form an integral addition to the sampling quality assurance procedures already noted.

4.3.8 *Sampling stages*

A sampling plan designed after site reconnaissance and 'desk study' should not be regarded as inviolate. It is likely to be altered by unforseen physical occurrences on site, e.g. the discovery of underground storage areas, encountering concrete foundations etc. and it should be altered in response to observation of strata, odour, appearance of deposits etc. It may also need to be reviewed when analytical data on the site samples is available. The sampling plan designed should be regarded as one which may need to be supplemented by a further stage of sampling. In some cases it may be prudent to design a first stage sampling/analysis plan which is an exploratory investigation needing to be followed by a second stage plan as a more in depth investigation. Such a two stage sampling plan may well be appropriate for example where site history is not well defined.

Acceptance that more than one stage of sampling can be desirable is dependent on having the time available for two or more stages of sampling and again emphasizes the need for early recognition of the need for site investigation.

In summary the following stages form part of an appropriately designed site sampling/analytical programme:

Recommended stages in site sampling/analysis

(1) Construct list of potential site contaminants
(2) Formulate broadly the extent of the sampling/analytical programme by considering the hazard associated with the potential contaminants in relation to the possible intentions for site development
(3) Plan detailed sampling programme which has:
    3.1 Surface sampling points related to observation made during the site reconnaissance
    3.2 Sampling points in and around potential 'hot-spots'
    3.3 A regular grid of sampling points, where grid size and numbers of sampling points is decided by:
        (a) knowledge of the site
        (b) reference to the approach described in sections 4.3.5.2, 4.3.5.3. and Table 4.6
        (c) reference to the approach described in section 4.3.5.4 and Table 4.7.
(4) Modify sampling plan in reaction to physical conditions on site or to observation during site sampling.
(5) After site sampling modify broadly defined analytical programme in reaction to site observations or any on-site testing effected.
(6) Consider results of analysis produced from (5) and decide whether adequate information has been gained to allow design of redevelopment programme—if not institute second stage of sampling/analysis.

Again it must be emphasized that a multi-disciplinary approach is essential to achieving a well planned and executed site investigation.

4.4 *On-site sampling, observation and testing*

There are four important areas for consideration:
Safety
Excavation methods
Sampling methods
On-site testing.

4.4.1 *Safety*

Before site sampling commences adequate measures must be made for the protection of those carrying out the on-site investigation. The nature of hazards will have been defined by the prospective survey and that information must be used to design protective measures in accordance with The Health and Safety at Work Act (30) and The Health and Safety (First Aid) Regulations (31).

Against the background of knowledge of the site, consideration must be given to the following:

(1) Protective wear for:

(1.1) Head

Since in most cases site work involves association with excavation machinery or drilling rigs, safety helmets should always be worn.

(1.2) Face

Where corrosive contaminants may be present, eye-protection will be appropriate, or in more extreme cases, full face visors might be required.

(1.3) Body

Disposable body suits are useful for many site investigations, but chemical resistant suits should be used especially where acid/alkali exposure is likely.

(1.4) Feet

Generally synthetic rubber safety boots will be appropriate.

(1.5) Hands

Disposable gloves should be the minimum requirement for those involved directly in fill or water sampling. In known corrosive conditions heavy duty synthetic rubber gloves should be worn.

(2) Respiratory equipment

Assessment of the need for respiratory protection will depend very much on the individual circumstances of site investigation.

The likely levels of exposure must be postulated and appropriate equipment sought by reference to the British Standard for selection, use, and maintenance of respiratory protective equipment (32) and to current Health and Safety Executive lists of approved respiratory protective equipment (33).

(3) Work in trenches

Concentration of vapour say, from petroleum hydrocarbons in a trench can be significant and extremely hazardous. On sites where gas or vapour release is thought likely, trenches should not be entered by the investigator, until these are adequately ventilated.

On other sites entry into trenches should always be made cautiously with a second investigator close by in case of emergency. Below 1.2 m depth entry must only be made after the trench has been structurally supported.

(4) Danger of fire
Vapours emitted from site contaminants may be ignited accidentally, e.g. by sparking from excavator movement. For example levels of around 4000 mg/kg of toluene and xylene (34) have been associated with small explosions and fires in contaminated surface fill. Site investigation machinery, e.g. excavators, should always carry a fire extinguisher, and a no smoking ban should apply to the site outside the mess cabin.

(5) Hazard monitoring
Where vapour or gas release may occur, site investigators should carry out regular periodic monitoring for example using grab-sampling detection tubes for solvent vapours.

Where the presence of particularly corrosive materials e.g. acids/alkalis is suspected then regular pH checks on fill/groundwater should be effected.

An experienced investigator familiar with the significance of the various exposure levels encountered should continue, modify or curtail the investigation in reaction to the measurements made.

(6) Hygiene
A site cabin should be provided for all investigations lasting more than a half day. The cabin must have facilities for changing, washing and eating. Eating and drinking on site (outside the cabin) should be prohibited.

Where extremely corrosive conditions may be encountered a facility should be provided for deluging exposed victim with water at the excavation point or site cabin as appropriate.

(7) First Aid
The following first aid materials should be available:

- (7.1) In the site cabin:
  A general first-aid kit for treatment of cuts etc.
- (7.2) In the site cabin and at the sampling area:
  Eye-wash bottles.

(8) Physical hazards
The normal awareness required in site investigation for engineering purposes is required (18) including for example prior location of electrical services.

4.4.2 *Excavation methods*

4.4.2.1 *Methods.* Methods of ground excavation for site investigation for engineering have been reviewed by Dumbleton and West (35) and similar methods are applicable to site investigation for contamination.

At the site surface excavation for sampling of surface fill is readily achieved using manual methods. Below say 0.5 m the choice lies between trial pit/trench excavation or borehole probing.

Trial pit/trench excavation (figure 4.4) is likely to play a part in all contaminated land investigations and apart from cases where:

Figure 4.4 In preparation for excavation of trial pits

(a) Sampling below around 6 m depth is required, or
(b) specialized water or gas sampling is required

then pit/trench excavation will be preferred to borehole probing.

The advantages of pit/trench excavation over borehole probing can be summarized as follows:

(1) The experienced investigator can observe the strata in the site and any unusual deposits with greater ease and efficiency.
(2) In reaction to observation, trial pits can be extended into trenches to investigate a particular occurrence.
(3) Obstructions (e.g. boulders, fragments of concrete) can be dealt with relatively easily and trial pit excavation resumed.
(4) Around 13 excavations (3 m depth) may be achieved per day.
(5) Groundwater located is more easily and cleanly sampled without sophisticated sampling devices.
(6) Overall, site investigation using excavation methods will be much more flexible and less costly than borehole probing.

The maximum depth of trial pit excavation (using a track driven excavator) will be around 6 m; wheel-driven vehicles can sample up to around 3.5 m depth. Since in many cases sampling will not be required below around 2.5 m the depth restriction is of little significance.

It should be noted that, in some unstable ground conditions (e.g. land—fill sites) pit or trench collapse may occur after say 2 m. This effectively prevents sampling at greater depth without shoring of the pit. In these particular circumstances it may be more cost effective to adopt borehole probing for sampling at greater than 2 m whilst retaining excavation to benefit from observing materials in the upper strata.

Apart from the situation described above, borehole probing will be needed for sampling at greater than the maximum excavation depth of 6 m, and for gas sampling or longer term groundwater sampling.

Generally borehole probing by rotary cutting tools is likely to be of little value since the material produced on the site surface will be too disturbed. Borehole probing by tube driving will be of greater value although again when observing materials on the surface the likelihood of disturbance must be recognized. Lubrication of the drilling tool may be deemed necessary and as far as possible this should be restricted to water rather than a hydrocarbon oil. Borehole probing for gas or water sampling generally has the aim of leaving in-situ some sample collection device which will allow periodic monitoring of gas emmission or of water quality. These techniques are referred to later.

4.4.2.2 *Information requirements and recording*. Since the purpose of ground excavation is to observe the site and sample its strata to provide information for design of the redevelopment it is vital that the excavations are properly supported by recorded plans and details of observations made and samples taken.

Guidance on information requirements and data recording is given in two British Standards Institute publications (7, 18) and by Dumbleton and West (35).

However in summary, each on-site investigation must be supported by the following documentation:

(1) Safety requirements
The detailed requirements to maintain the safety and well being of the site investigators must be documented, and read and understood by each of those involved. The document should act as check-list of protective equipment etc. needed for the investigation.

A typical site investigation team might consist of chemist, and environmental health officer or engineer, together with an excavator driver. It is important to exclude the driver from any contact with the sampling operation (i.e. other than operation of the excavator), so that the most stringent protective equipment requirements apply only to the chemist and his associate.

(2) Sampling plan
The previously determined grid and hot-spot sampling point locations should be marked on a suitable plan for use by the investigators. Locations will have been previously marked by the engineer. New locations chosen during the sampling exercise should be clearly marked for subsequent measurement and plotting on a revised plan.

(3) Sampling point and sample codings
Each sampling point location must be clearly and unequivocally coded and each sample taken must be coded in relation to location and depth of sample to avoid confusion on site and subsequently in the laboratory between samples, locations and indeed sites.

(4) Recording of sampling information
A logging system should be used so that for each sampling point the following is noted:

(a) Description of site surface
(b) Total pit depth, or depth and length in case of trench excavations
(c) Appearance and depth of strata observed
(d) Appearance, odour and depth of any unusual deposits
(e) Appearance, depth and relation to strata descriptions of samples taken
(f) Any other observations relating to odour, appearance, and groundwater occurrence and depth.

A useful aid to subsequent discussions of the redevelopment options for the site are a series of photographs indicating the site (or parts of the site) as a whole and particular strata or pockets of visible contamination.

4.4.3 *Sampling methods*

4.4.3.1 *Requirements of sampling method.* The method adopted for sampling soils, waters, gases etc., from a contaminated site must guarantee:

(a) That material of interest is sampled
(b) That the sample remains stable until analyzed
(c) That the sample does not undergo contamination before being analyzed, and
(d) That the state of the sample is compatible with the method of analysis to be used.

Fulfilment of condition (a) above may be partly achieved for example by taking fill samples at pre-determined depth intervals, but is also concerned with the recognition of fills that are usefully sampled. The experienced investigator can significantly improve the ability of the sampling programme to detect contamination, by taking samples in reaction to the appearance and odour of the fill.

To ensure that the sample remains stable and that the sampling procedure is compatible with the method of analysis, again demands the multidisciplinary approach recommended earlier. Sampling carried out by one body in isolation from its analytical laboratories could produce a false picture of the site by using inappropriate sampling methods for example which, say, allowed the loss of volatile contaminants.

4.4.3.2. *Sample stability*. The stability of site samples has been discussed in the proposed BS Code of Practice (7), by Wilson *et al.* (9) and by Barth *et al.* (27) and the stability of water samples by the DoE/NWC Standing Committee of Analysts (25).

Broadly, stability of samples can be considered in three classes:

(1) Samples in which the contaminant(s) are stable.
(2) Samples in which the contaminant(s) are unstable but stability can be achieved by a preservation method.
(3) Samples in which the contaminant(s) are unstable and cannot be readily stabilized.

For those contaminants which are unstable then loss or change (chemical or biological) of the contaminant must be minimized by either preserving the contaminant, e.g. freezing or addition of a stabilizing agent or by arranging for analysis to be undertaken immediately or soon after sampling.

Examples of preservation systems are given in references (9) and (25) for a wide range of inorganic and organic contaminants.

Some contaminants are not easily stabilized, at least in a manner compatible with subsequent analysis. Volatile solvents fall into this category and perhaps present the greatest difficulty in sampling, furthermore literature guidance on preservation is sparse. Fill contaminated by say 500 mg/kg of toluene may begin to lose the hydrocarbon by volatilization as soon as the

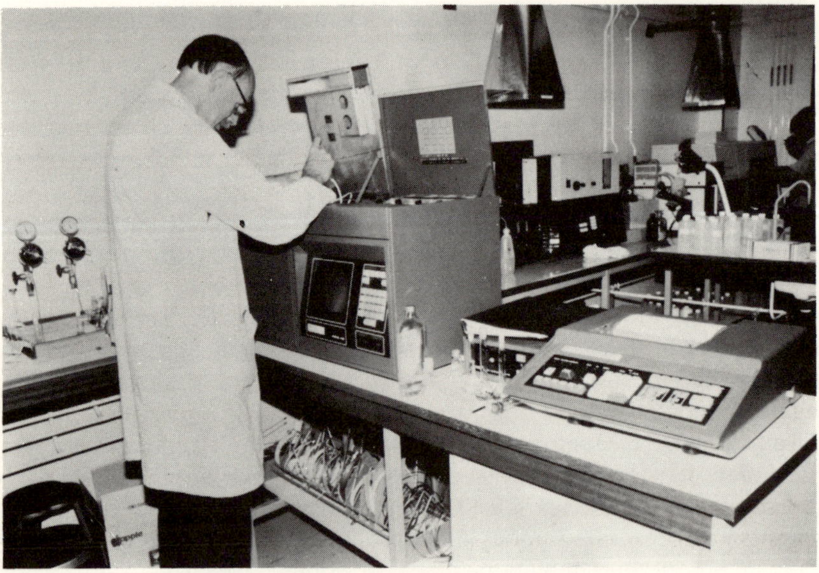

Figure 4.5 Determination of solvent contamination by gas-liquid chromatography

contaminated fill is exposed by excavation. A sampling procedure is needed to minimize such loss.

Four potentially useful approaches to avoid underestimation of solvent concentration from the sample taken are:

(1) Analysis of material on site immediately after sampling
(2) Transport of material to laboratory to enable analysis to be started within hours of sampling
(3) Sample materials into a non-volatile compatible solvent
(4) Contain samples in a gas-tight vessel and store under freezing conditions until commencement of analysis within 24–48 hours.

In the paper by Aldred and Lord (8) two solvent contaminated sites are referred to. They were sites contaminated by aliphatic and aromatic hydrocarbons, dichloromethane and trichloromethane. Sampling was effected by combinations of procedures (2), (3) and (4) above.

It was found by 'spiking' experiments that recoveries were variable and dependent on time lapse between sampling and analysis, storage conditions, nature of soil and nature of solvent. Recoveries ranged from 50% to 90% and no firm conclusions were reached on sampling method other than the need to assess the sampling procedure adopted, by a series of 'spiking' experiments until recovery is satisfactorily estimated.

One consequence of sampling schemes adopted for solvent measurement is that since sample homogenization cannot be effected without loss of solvent, then the sample weight taken is that used in analysis and is generally small, say 50 to 100 g.

Thus when a stratum of unusual material is observed, a sample taken for solvent examination, say of 50 g may be less representative of the stratum than one taken for say lead analysis of 1000 g. It will therefore usually be necessary to take more than one sample for solvent examination to achieve better representation of solvent concentration at any point and for the site as a whole.

Once sampled into its container a sample could become further contaminated if:

(a) The container was not secure
(b) The contaminants were aggressive towards the container.

For water samples, reference (25) offers information on containers for aggressive materials but generally the three most useful forms of container for fill and water samples are polyethylene bags; polyethylene bottles/boxes; glass jars. Avoiding container insecurity or breakage is largely a matter of good organization and careful attention particularly when the more vulnerable bags are used for fill sampling.

4.4.3.3. *Summary of sampling practices.* (a) Sampling of soil/fill: Stainless

Figure 4.6 Sampling of gas from a borehole

steel or polypropylene trowels can be used to sample directly from the trench (i.e. entry only to 1.2 m depth) or from the bucket of the excavator.

Where a stratum of material is observed sub-samples should be taken across the stratum to achieve as representative a sample as possible.

The weight to be taken will be subject to analytical method requirements, but in many cases a sample weight of around 1 kg will be appropriate.

(b) Sampling of water: Samples from small areas of surface water can scooped (polypropylene) or collected by vacuum pipette, avoiding the ingress of solids.

Groundwater issuing from a point or stratum into a trial pit should preferably be sampled at the point of issue by collection into a polypropylene vessel for transfer to the sampling container.

If sampled from the base of the trench or excavator bucket, solid ingress will be difficult to avoid, and the sampling conditions should be noted on the trial pit log.

Where larger areas of water e.g. ponds, lagoons are to be sampled along with their associated sediments, advice on appropriate methods is given in Reference 25.

For some site investigations, examination of groundwater quality over an extended period of time may be appropriate. To meet that requirement, periodic sampling of water sampling boreholes may be needed.

(c) Sampling of gases/vapours: Gases/vapours may be need to be sampled and analyzed to:

(a) Indicate their presence or absence in a site investigation, or
(b) measure their emission rate from a site.

In both cases sampling (Figure 4.6.) may be effected by insertion of a sampling pipe into the ground, allowing the gases to collect in the pipe and come to equilibrium with the 'reservoir' below ground, then taking a sample of the equilibrated gas for analysis. To measure emission rate as opposed to simply indicating presence, monitoring will be required over an extended period of time. In addition the sampling pipe may be flushed free of gas, using say nitrogen, and the subsequent increase and maximum concentration of gas measured.

The most common application of borehole gas sampling is in the measurement of methane, carbon dioxide and others from landfill sites.

Methods of sampling gases from landfill sites have been described by Thorburn *et al.* (36), Griesinger (37), McBean and Farquhar (38) and Aldred and Lord (8).

Methane sampling is considered in detail in Chapter 11.

(d) Sampling of plants and grasses

It may be pertinent as part of site investigation to sample root crops, leaf crops or grasses. Root crops will already be contaminated by adhering soil which will need to be removed before analysis. Leaf crops and grasses need to be sampled and packaged carefully to firstly avoid further contamination and secondly to present suitable material for any subsequent washing needed before analysis.

Advice on sampling procedures for such mediums is given in References 39, 40 and 41.

4.4.4. *On-site testing*

On-site testing facilities may be required:

(a) To support health and safety remedial measures by indicating the presence say of toxic vapours or acidic liquids etc.
(b) Because immediate measurement of contaminant is the only way to quantify its concentration i.e. because of its instability, volatility etc.
(c) To speed up analysis of site materials i.e. by having immediate analytical response
(d) To reduce the costs of analysis i.e. by avoiding transport costs.

A wide range of portable analytical equipment exists, an indicative list of some of the equipment more relevant to site investigation and readily available is shown in Table 4.8.

The main advantage of equipment of the types listed in Table 4.8 is that a truly portable system of analysis is made available. The two major disadvantages are likely to be:

Table 4.8

Equipment	*Uses*
Detector tubes e.g. Draeger, Gastec	Wide range of gases and vapours
Colorimetric field test kits available for water e.g. cyanide, phenol (Note 1)	Wide range of test kits
Dedicated gas monitoring instruments	Range available includes methane, sulphur dioxide, carbon monoxide, hydrogen sulphide
Dispersive infra-red analyzer (Note 2)	Measurement of a wide range of gases and vapours
Flame ionization (Note 2)	Measurement of wide range of
Gas chromatograph	volatile materials
Miscellaneous e.g. pH water, qualitative test kits, conductivity water.	

Note 1: Soil test kits are not generally available, but the possibility exists of extracting soils with water to give a test medium.
Note: 2 Capable of analysis of mixed component samples.

(1) The lack of specificity of some of the techniques indicated
(2) The lack of accuracy of some of the techniques.

Where such problems exist they may still be capable of resolution on site by establishing more sophisticated techniques of analysis in a temporary site laboratory or a mobile laboratory.

The use of mobile laboratories in the United Kingdom in contaminated site investigation appears to have been mainly restricted to gas and asbestos analysis. In the United States of America however much more comprehensive mobile facilities have been established which can provide facilities for rapid and accurate on-site assessment of pesticides, PCBs, phenols, metals etc. These facilities have been used on many different types of site investigation and some are described in References (42, 43, 44).

The establishment of a mobile laboratory with facilities matched to different site investigations, whilst capable of providing rapid response, will be expensive.

Whether the use of portable analytical equipment or of a mobile laboratory is justified will depend upon many factors. Certainly where safety measurements are needed (see earlier) or where the contaminant must be analyzed immediately if it is to be accurately quantified, then some form of on-site testing facility is needed. But in other cases cost, speed, convenience etc., have to be assessed in the knowledge that in the United Kingdom an established analytical laboratory is likely to be relatively close to a site or to be able to offer a comprehensive sampling/analytical facility which may obviate the need for on-site testing (i.e. other than in the cases noted above). For example the

United Kingdom's Public Analysts have their 51 laboratories geographically spread to provide an analytical service to all parts of the United Kingdom; in addition the Royal Society of Chemistry (45) lists consultant laboratories whose spread is also wide.

For a detailed review of on-site testing methods the reader is referred to Reference 46.

4.5 Analysis of material taken from the site

4.5.1 *Sub-sampling in the laboratory*

4.5.1.1 *Control of sub-sampling procedure.* Some materials will be received in the laboratory needing no further sample preparation and ready for analysis, e.g. air samples contained in gas-tight flasks. Others will require preparation prior to analysis.

The method of preparation must be carefully chosen to:

(a) Avoid bias in the analytical result e.g. by 'driving off' a volatile material in a drying stage.
(b) To contribute a minimal level of random error in the analytical result e.g. by grinding down to the particle size that produces the optimum precision in the data.

The elimination of bias in the sub-sampling procedure is largely a matter of considering each of the steps in the procedure, in the light of the form in which the contaminant exists. Where there is doubt that a particular stage might result in loss or gain of the contaminant, then the procedure should be carried out with a suitable matrix to which has been added a known concentration of the contaminant in question (see for example the solvent sampling methods described earlier). Once the sub-sampling procedure is chosen, it is essential to control possible bias in day to day use of the method. This can be effected by:

(a) Properly documenting the procedure
(b) Standardizing the equipment used
(c) Adequately training the users of the method and making them aware of the consequences of changes in procedure

4.5.1.2 *Particle size.* Since a soil/fill sample will be a heterogeneous mixture of different materials and particle sizes its preparation will necessitate mixing and grinding. The greatest influence of the preparation procedure on the precision of analytical results will be the particle size to which the material is ground.

Probably the most common grinding procedure used in soil/fill analysis is that recommended by ADAS (41) where the soil is ground to pass through a 2 mm sieve. The 2 mm is considered to be a typical soil crumb size and to be most pertinent to soil nutrient analysis. The same 2 mm size was used in the study of cadmium soil levels at Shipham (47) and is recommended for study of gasworks sites by Wilson *et al* (9).

The proposed BS Code of Practice (7) recommended grinding to pass through a 1 mm sieve. The Standing Committee of Analysts (39) recommends a procedure involving grinding the whole sample to pass through a 2 mm sieve. But for 'special determinations and when it is necessary to weigh amounts of less than 1 g' they recommend that the whole of the <2 mm sample should be quartered, cored and a sub-sample of 20 g ground to pass a 0.5 mm sieve.

Smith and Ellis (23) examined the difference in grinding soils from a gasworks site to:

(a) 'About 1 mm'
(b) 'About 150 μm'

by analyzing soils, so ground, for total sulphur and acid-soluble sulphate content. Using the same method of analysis a series of samples were analyzed in triplicate and the coefficient of variation calculated for each series of triplicate analyses. Their results are shown in Tables 4.9 and 4.10.

Clearly the data in Tables 4.9 and 4.10 shows a general trend towards significant improvement in precision when particle size was reduced from 'about 1mm' to 'about 150 micrometre' for these specific analyses.

In the author's laboratory, samples are generally ground in a hammer mill to produce particles less than 1 mm.

The difference in precision achieved from that procedure and further grinding to produce material passing through a 75 micrometre sieve has been compared.

The comparison was made using a method for determination of total lead which involved taking a 1 g sub-sample of the ground material, ashing at 450 °C, extracting the lead using nitric acid and measurement by Atomic Absorption Spectrophotometry (Flame mode).

Table 4.9 Total sulphur data from reference 23

Sample reference	*Ground to 'about 1 mm'*	*Coefficient of variation % Ground to 'about 150 micrometre'*
B	8	1
C	9	2
D	4	1
E	10	10
F	15	2

Table 4.10 Acid soluble sulphate data from Reference 23

Sample reference	*Ground to 'about 1 mm'*	*Coefficient of variation Ground to 'about 150 μm'*
A	6	2
B	5	5
C	19	8
D	51	23
E	22	15
F	4	1

Table 4.11 shows the results of experimental work where the results normally obtained by the author's laboratory are compared to those obtained when further grounding to less than 75 micrometre is carried out. The results are also compared with the results of analysis of sub-samples from a soil reference material produced by the Community Bureau of Reference (48). The latter results are assumed to indicate the precision of the analytical method itself.

The data is expressed in terms of coefficient of variation (cv) which is standard deviation expressed as a percentage of the mean value of the results.

The table shows that grinding to 75 micrometre produces a precision similar to that obtained using the reference material, and in that sense could be regarded as the optimum precision.

Grinding to less than 1 mm gives poorer precision but in the case of this particular analysis will in most cases be adequate for interpretation of site data.

The coefficient of variation of 2% achieving by grinding to less than 75 micrometre is comparable to that of 2% achieved in the author's laboratory for the determination of chloride in concrete. The relevant British Standard specifies grinding of the concrete to less than 150 micrometres prior to analysis (49).

Carlton-Smith *et al.* (50) describe an evaluation of the homogeneity of soil which was 'dried in a draught-oven not exceeding 30°C and ground to pass a 0.5 mm sieve. It was then mixed thoroughly until heterogeneity reached a minimum'.

Table 4.11

Sample	*Coefficient of variation of results*
Ground to less than 1 mm	13
Ground to pass 75 micrometre sieve	2
Reference material	2.5

Homogeneity was checked by taking sub-samples from random locations within the ground sample and comparing the subsequent results of analysis with those from determinations on one sub-sample from the sample (to represent the variance inherent in the method). Analyses were performed by atomic absorption spectrophotometry and x-ray fluorescence spectroscopy. Carlton–Smith *et al.* concluded that the heterogeneity of the sample accounted for increases in coefficient of variation of:
$> 1\%$ (absolute) for cadmium, chromium, copper and zinc
and 2% (absolute) for nickel and lead.

The soil prepared by the procedure described was deemed by Carlton-Smith *et al.* to be of adequate homogeneity for a collaborative test of a method of analysis.

It is apparent that grinding to a smaller and smaller particle size will facilitate the production of analytical data with better and better precision. It would seem that grinding to 150 micrometres or less might be expected with some analytical parameters to show coefficients of variation around 2%. It is however unlikely in most site investigations that such precision will be needed, bearing in mind the errors associated with sampling and with interpretation of data (see section 4.5.2.3).

It is the author's view that as a general rule grinding to less than 1mm should be adopted as primary practice with recourse to smaller particle size where improved precision is needed.

Since the overall method of analysis (i.e. including the sub-sampling procedure) will be properly validated and subject to quality control (see section 4.5.3) then any major deficiency in the precision of data will be identified.

4.5.1.3 *Detailed procedures*. The detailed sub-sampling procedure adopted in the laboratory will depend also on the nature of the contaminant. Reference has been made already to the effect of stability of contaminants on sampling procedures on site and stability equally effects sub-sampling protocols within the laboratory.

Sub-sampling procedures are detailed in the proposed BS Code of Practice (7); the author's modifications of those described are:

(a) For stable contaminants in soil/fill

Step 1 Empty the 1 kg sample on to a clean surface, spread to examine. Remove material greater than 2 cm maximum dimension, which is not pertinent to the analysis, e.g. stones, bricks. Break down other material by hand until less than around 2 cm. Mix thoroughly.

Step 2 Turn on to clean tray and dry at not greater than 30 °C in a draught-oven.
Grind the whole sample, until it has a particle size of less than 1 mm. Mix throughly and transfer to a well sealed container.

(b) For unstable contaminants in soil/fill

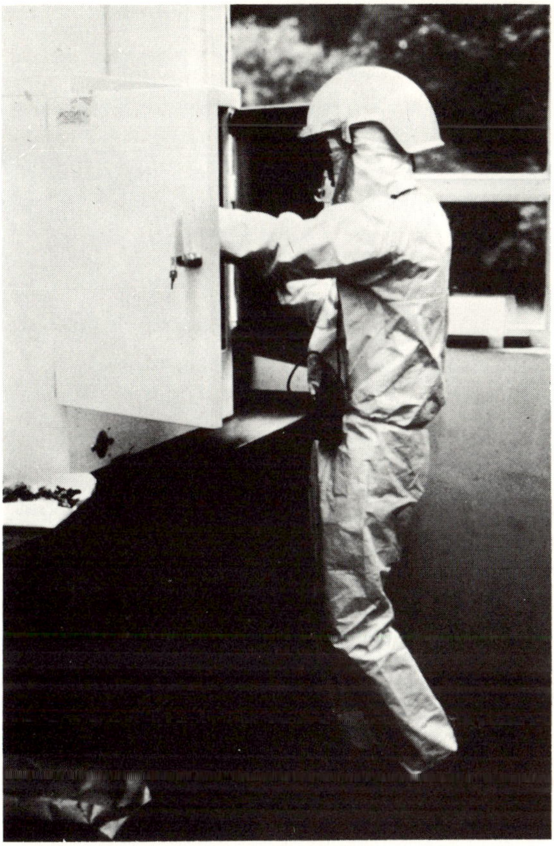

Figure 4.7 During drying and grinding of soils in the laboratory, safety measures need to reflect the hazard to laboratory staff—an extreme example

If the contaminant is capable of being stabilized by addition of a stabilizing agent but this has not been effected on site, then Step 1 above should be completed quickly and a sub-sample taken to a suitable container for the addition of the stabilizing medium.

For contaminants not capable of being stabilized, Step 1 should be completed quickly and a sub-sample analyzed immediately.

The latter two procedures can be expected to produce analytical results of poorer precision because of the lack of homogeneity of the sample. In these cases, improvement in the precision of data can only be effected by increasing the number of analyses carried out.

(c) Sub-sampling of water:

Any organic phase must be separated and retained for analysis. Insoluble matter should be filtered (Whatman GFC or equivalent) and discarded.

Recommended stabilizing agents, storage temperatures etc. are given in Reference 25.

(d) Sub-sampling of grasses and plants:

Samples requiring washing to remove surface contamination should be well separated and treated accordingly.

Prior to analysis all samples should be well mixed, dried at 30 °C in a draught oven, chopped finely and then ground to particles of less than 1 mm.

4.5.2 *Selection of analytical methods*

4.5.2.1 *General principles.* The general principles of selecting an analytical method are those pertaining to other branches of analytical chemistry. The particular significance of some of those principles in the case of analysis for site investigation is considered below.

The method selected must meet the following criteria:

(i) Be capable of producing results of the required accuracy
(ii) The effect of likely interferences on the ability of the method to produce accurate results must be understood
(iii) Have had its validity established by sufficient testing
(iv) The time taken for analysis by the method should be such as to meet the time requirements of site investigation
(v) The time, equipment, materials etc., associated with the method should be such as to present the optimum cost at the performance level required
(vi) It must measure the criterion of interest
(vii) It must have a detection limit appropriate to the problem.

The problem of interference in contaminated site investigation may be significant because of the presence of many contaminants other than the parameter of interest or because of the differences in matrix effects within one medium (e.g. different types of soil/fill) or because of the desire to apply one method to a number of different mediums e.g. soil, water, plants. Clearly, validation of the method must take into account potential interference effects and there must be an adequate quality control procedure (see later) to check on the continued performance of the method.

4.5.2.2 *Nature of the analytical parameter.* For many analytical parameters consideration needs to be given to the form in which the parameter is determined. Particularly, for example, in the case of metals analysis, the parameter may be estimated as

Total
Water-soluble
Acid-soluble, or
Plant-available

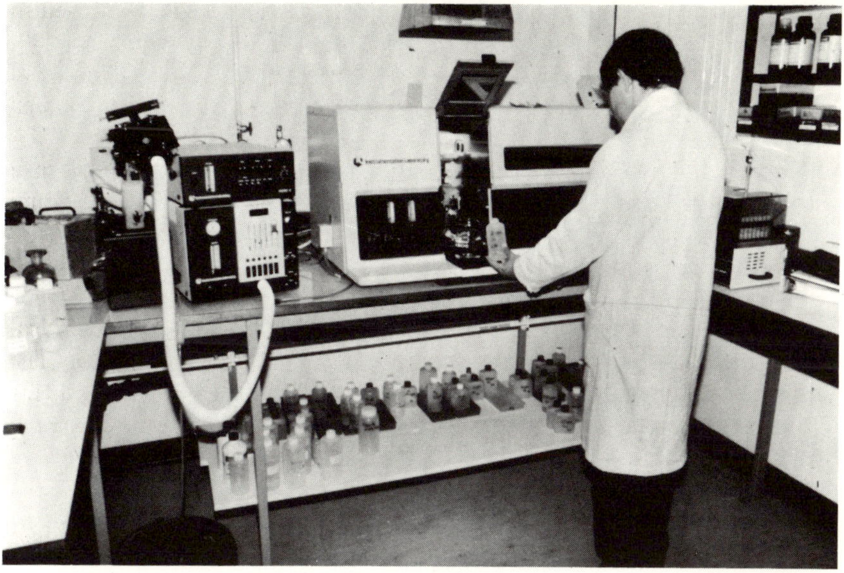

Figure 4.8 Determination of metallic contaminants by atomic absorption spectrophotometry

The descriptions signify the manner in which the soil is extracted for determination of the metal.

'Total' metal determinations are commonly effected by decomposing the matrix by wet oxidation or dry oxidation.

Wet oxidation, i.e. with acids, has the advantage of being effective on organic and inorganic materials, and can reduce interferences by efficient matrix destruction. In addition, loss of volatile metals is prevented by the relatively low (say 100 to 200 °C) oxidation temperatures. However the technique can be lengthy, prone to contamination, and requires greater expertise than dry oxidation.

In dry oxidation, the sample is heated (usually to not greater than 450 °C) in a furnace to destroy the organic matrix. The residue is dissolved in acid. The technique is fast, less easily prone to contamination but can result in loss of volatile metals.

For determination of total lead, cadmium, chromium, copper, nickel and zinc in sewage sludge, the DoE/NWC Standing Committee of Analysts (51) has recommended a simple digestion technique in which the sample is heated for 15 minutes in nitric acid. The DoE/NWC (52) also state that the technique will not release the total metals in samples containing a siliceous matrix, and in those circumstances wet oxidation with nitric/perchloric/hydrochloric acids may be required. In the later case, though, it is also stressed that 'elements firmly bound in a siliceous matrix are not normally of physiological significance'.

Methods for wet oxidation of soil by nitric/perchloric acids for determination of lead, cadmium, copper and others are prescribed by ADAS (41, 54).

In general wet oxidation is to be preferred for total metal determination, but dry oxidation can give adequate recovery of metals and may be useful in some matrices where large numbers of analyses are required quickly.

In the author's laboratory a dry oxidation technique followed by nitric acid dissolution has been found to give recoveries of lead and cadmium within referencc specification using Community Bureaux of Reference soil reference specimens (the use of reference specimens is discussed later in the chapter).

Plant available metals are determined (41, 53, 54) using extractants which empirically assess the availability of the metallic soil constituent to the plant. Two extractants are most widely used, acetic acid and ethylene diaminotetra-acetic acid (or its salts) (EDTA).

The use of acetic acid reflects to some extent the acidic nature of many United Kingdom top soils and the EDTA extractant reflects to some extent the potential effect of calcareous soils containing free lime.

Analytical data produced using such extractants are empirical—their interpretation is discussed in reference 54 and is the subject of a DoE Working Party Report (55).

Acid-soluble metals (i.e. without wet oxidation) in soils from contaminated sites are sometimes determined. Such extractions have been claimed to be an indication of human availability of metals from ingested soil since gut conditions are acidic. One extractant used has been 0.07 N hydrochloric acid which is the extraction medium specified in certain consumer protection legislation (56), where lead and other metals extracted from paint films etc., are determined.

At first sight the possibility of using an analytical method which produces results directly related to hazard is attractive. However little work appears to have been carried out on the relationship between acid-extractable metal levels and influencing factors such as the nature of the soil, the strength of the acid, the time of extraction etc. Furthermore even the interpretation of the significance of acid soluble metals by a 'standardized' procedure is likely to be no less difficult than interpretation of total metal levels.

The information of contaminants needed for design of a covering system may be restricted in many cases to nature of contaminant, an indication of its potential effects, concentration in soil and concentration in groundwater.

In cases where sufficient groundwater contamination data is not available, or may not adequately represent the long term prospects for contaminant mobility, then further information on contaminant solubility in water may be required.

As with acid extraction procedures water solubility determinations are dependant on extraction time, extraction temperature, soil type, particle size of sample, composition of extractant and other factors.

Almost invariably, assessment of 'total', 'acid soluble' etc., data will be assisted by comparison with data from other locations, contaminated or uncontaminated. Those comparisons can only be valid if the concepts of 'total' etc., are the same, thus care must be taken when comparing data which is not produced by identical methods of extraction.

It is particularly important to measure by the same defined procedure when an indicator method is used. An indicator method measures, as the name suggests, a general property of the sample rather than a specific contaminant. For example the 'toluene extract' of soils has been widely determined on former coal gas production sites as an indicator of coal tar contamination. That and similar methods may be valid primarily because of their low cost operation, but comparison with similar data must be procedurally consistent, and hence must only be made when the method of test is precisely the same.

4.5.2.3 *Accuracy of results.* The major factor influencing the choice of method of analysis in any field of analytical chemistry is the accuracy required of the results.

In many areas of analytical chemistry precise standards or guidelines exist e.g. United Kingdom Regulatory standards; British Standards; Codes of Practice; EC Directives etc., which form a basis for prescribing the accuracy required of results.

In the area of site investigation there are no United Kingdom prescribed standards, but the following will influence the decision on the required accuracy:

(i) Existence of DoE 'trigger concentrations' for comparative purposes (see later)
(ii) Existence of other guideline values (see later) for comparative purposes
(iii) Existence of data on 'background' or 'normal' levels in soil for comparative purposes (see later)
(iv) Existence of data relating contaminant concentration and exposure to effect on health
(v) The relationship between the data and design of remedial works
(vi) The relationship between accuracy, cost, speed and overall resources available.

The multi-disciplinary approach advocated previously can create the conditions under which discussion between the analyst and his fellow site investigators evolve so that the optimum selection of analytical method is made by the analyst.

The proposed BS Code of Practice (7) did observe that the error associated with results of analysis must be seen in perspective to the total errors associated with site investigation.

Undoubtedly the site sampling programme is likely to be associated with the greatest error in evaluation of site conditions.

Furthermore, interpretation of analytical data in terms of health risk, risk to building materials, design of remedial works is unlikely to be based on accurately defined guideline levels. Therefore error associated with data interpretation must also be seen as a significant part of the total error concerned with site evaluation. By comparison errors associated with laboratory sub-sampling and analysis are likely to be small.

The analyst must seek to estimate and control the errors involved in his laboratory sub-sampling and analytical procedures but in specifying the accuracy of his methods he must bear in mind their relationship to the total error associated with site investigation.

4.5.3 *Analytical quality assurance*

There is not a United Kingdon catalogue of officially recommended methods of analysis specifically for contaminated site investigation, and it is likely that many different methods or variations of methods are used in United Kingdom laboratories concerned with site investigation. It would be of great benefit to those laboratories if a series of published recommended methods existed similar to the series produced by the DoE/NWC Standing Committee of Analysts mainly for water analysis. Certain method of analysis in that series are recommended as suitable for use in sewage sludges, sludge treated soils and sediments and as such are likely to be applicable to contaminated site investigations—this is considered later.

That series presents a group of methods of analysis whose performance characteristics have been well defined.

Performance characteristics (57, 58, 59, 60) are as the name suggests quantitative parameters which define how the method performs and define the accuracy of the results produced.

The term accuracy is used in this section to define the total error associated with an analytical result. Where total error is a combination of systematic error and random error (61). The terms bias and precision are used respectively to define systematic error and random error.

The DoE/NWC series (25, 62) of methods generally describe the following performance characteristics:

(i) The applicability of the method in terms of the nature of the substance determined, the types of sample suitable and the general basis of the method
(ii) The concentration range over which the method can be applied
(iii) The nature of the calibration curve
(iv) A measure of the precision—generally a parameter referred to as the total standard deviation is quoted. This is a measure of two sources of random error—within and between batches of analysis
(v) The limit of detection of the method—often quoted from the calculation

$$2\sqrt{2}.s.t$$

where s = estimate of within-batch standard deviation of the blank determination
t = value of Student's 't' at 0.05 probability level

(vi) Sensitivity of the method—defined as the rate of change of analytical response with respect to concentration

(vii) Bias

Recognizing the difficulty of prescribing a quantitative value to the bias associated with a result of analysis the DoE/NWC series deals with bias in its component parts data on which is given particularly in (i), (iii) above and (viii) below

Further data for example on the instability of samples, comparative methods of analysis and general sources of error may also be given

(viii) Interferences—information on interfering and non-interfering compounds is given

(ix) The time required for analysis is indicated

(x) Lastly the methods often indicate a suitable method of checking the accuracy of results i.e. a quality control procedure although the frequency of checks required is not generally indicated.

Where the DoE/NWC methods are appropriate for contaminated site investigation work, then the analyst can determine the performance of the method in his laboratory and compare his performance with that described for the method. Such verification work comes within the general description—Analytical Quality Assurance (AQA).

AQA procedures are discussed by the DoE/NWC Standing Committee of Analysts in Reference 25.

Quality assurance generally in environmental chemistry is described by the American Chemical Society's Committee on Environmental Improvement in Reference 24.

Again, mainly in the field of water analysis, AQA protocols are defined by the US Environmental Protection Agency in Reference 63.

Perhaps the world's best established AQA procedures— those of the American Association of Official Analytical Chemists are described in Reference 64.

A useful definition of AQA has been suggested by the UK's Association of Public Analysts (APA) (65):

> 'The internal and external systems which evaluate a laboratory's performance and provide for a continuing process of monitoring that performance'.

With respect to internal AQA, the APA has recommended to its member laboratories that where performance data is not associated with official or recommended methods of analysis, then verification work (in all their various fields of activities) should be standardized to determine (within each individual laboratory) a minimum of:

(1) Precision—defined as within laboratory repeatability precision (W_p).

W_p is a concept related to replication experiments and is the maximum absolute difference (95% probability) between two analytical results obtained by the same method in one laboratory. W_p is applicable to analytical results produced by different analysts on different days using different reagents etc. and is calculated as:

$$W_p = t\sqrt{2}S_p$$

where S_p = standard deviation of replicates
t = critical value of 't' (95% probability)

(2) Bias
The APA recommends that two method of estimating bias be used:
(a) Analysis of reference materials
(b) Determination of recovery by spiking experiments.

Reference materials generally are available from the Community Bureau of Reference (48) and from the National Physical Laboratory (66). However the only materials directly applicable to contaminated site investigation at present available are a series of soils from the Community Bureau of Reference with known levels of lead, cadmium, zinc and other metals.

Thus with acceptance of the limitations of the technique, spiking of the matrix being tested with the substance (or similar) of interest is recommended as a useful indicator of bias. The APA protocol requires estimation of the within laboratory recovery mean value and within laboratory recovery standard deviation from a series of recovery experiments using the same method in one laboratory by different analysts on different days using different reagents etc.

(3) Limit of detection
Calculated from $tS_b + tS_s$
where S_b = standard deviation calculated from blank determinations
S_s = standard deviation calculated from sample determinations at a level close to the blank.

(4) Lastly the APA protocol for internal verification describes quality control procedures to be used as a continual check on performance.
Quality control checks are required at the rate of not less than one per 10 analyses by:

(a) Measurement of W_p above, or
(b) Spiking experiment, or
(c) Analysis of reference material, or

a combination of the three.

The internal quality control rate of not less than 10% (i.e. one in 10) is the same as that commonly used in United States Environmental Protection Agency recommended methods for water and wastes analysis (67).

An essential part of any AQA system is a system of interlaboratory collaborative testing. The APA external protocol embraces an organized system of collaborative trials but to date all of these have been in fields of activity other than contaminated site investigation.

Some AQA procedures in reported site investigations are reviewed below:

Aldred and Lord (8) report the use of the Association of Public Analysts protocol described above.

Brown and Black (68) describe in detail procedures used in Love Canal and Dallas lead in soil investigations. They developed a protocol whose objective they state was 'to provide monitoring data that possessed the maximum accuracy attainable'. The Brown and Black procedures and their frequency of use are shown in Table 4.12.

Performance requirements for each of the 12 procedures shown in Table 4.12 were laid down by the authors. For example in Procedure 7, the recovery experiments were expected to lie within 90–110%; in Procedure 9 the reference method of analysis should produce results agreeing within 10%.

It was the objective of the procedure to provide the 'maximum accuracy attainable' and that may not always be the objective of the analyst. For the reasons given earlier less accurate data may be quite appropriate for site investigation. Nevertheless the Brown and Black protocol offers useful guidance to the analyst.

Di Domenico *et al.* (69) describe the analysis of soils, vegetation and

Table 4.12

	Procedure	*Frequency*
1.	'Field blanks'—analysis of water which was used to wash the sampling equipment.	1 in day
2.	Laboratory water blanks—samples of water used to wash sub-sampling equipment in laboratory.	1 in 40 samples
3.	Duplicate samples—additional soil sample collected at the designated sampling location	1 in 20 "
4.	Triplicate sub-sample—a sample divided into three equal portions in the laboratory and each portion analyzed.	1 in 20 "
5.	'Calibration control standard'—repeat analysis of standard used for calibration.	1 in 20 "
6.	Analysis of reagent blanks.	1 in 20 "
7.	Recovery experiments—by spiking a sample extract with lead nitrate. *Notes*—Spiking level equal to endogenous level or 10 × instrument detection limit.	1 in 20 "
8.	Re-extraction of samples (i.e. to determine extraction efficiency).	1 in 20 "
9.	'Total recoverable analysis'—analysis of samples by a referee method.	1 in 40 "
10.	'Laboratory Control Standards'—analysis of reference material.	1 in 20 "
11.	'Split extract analysis'—analysis of second portion of samples extract.	1 in 20 "
12.	'Spiked sample analysis' (of 7 above)—Duplicate portion of sample spiked with lead.	1 in 20 "

Table 4.13

Procedure	*Frequency*
1. Reagent blanks.	'Periodic'
2. Identity checks on standards by second analytical technique.	Not given
3. Surface spiking—various surfaces e.g. stone, marble were sampled by a 'wipe test'; its efficiency was checked by spiking surfaces with solutions of TCDD.	28 spike tests
4. Tests to ascertain the linear response range of glc/ms systems and minimum TCDD amount capable of producing a recognizable signal.	Daily or weekly
5. Ratio of standard to sample injections.	Ratio mainly 1 to 2
6. Laboratory recovery tests—spiking with TCDD (a) to check whole method and (b) to check different clean up stages of method	Initial validation work plus continuing work at rate of 1 10 to 1 in 20 of samples
7. 'Test' samples subject to collaborative testing in up to 6 laboratories using same analytical procedure but different instrumentation	83 'test' samples used

miscellaneous surfaces for 2, 3, 7, 8—tetrachlorodibenzo-p-dioxin (TCDD) at Seveso, Italy. They required 'fast and reliable methods ... to ensure good detection thresholds notwithstanding sample matrix chemical complexity'.

The quality check tests adopted by the authors in a study analysing around 7000 samples are shown in Table 4.13.

The procedures shown in the references reviewed indicate the practical application of the AQA methods discussed earlier.

The precise form of a quality control procedure to be used in an individual laboratory will depend upon a number of factors and will vary from method to method.

However as a guide, the following is seen by the author as a useful basis for many quality control procedures:

(i) A standard of appropriate concentration should be passed through all parts of the analytical procedure at a rate equivalent to not less than once per batch of analyses or a rate of 1 in 20 whichever is the greatest.

This procedure facilitates control of all reagents used, checks calibration and is indicative of procedural bias.

(ii) Not less than 1 in 10 samples should be subject to 'spiking'. Analysis of the spiked sample should give results conforming to pre-determined performance characteristics.

(where a suitable reference material exists this may be substituted for the analysis of the 'spiked' sample).

(iii) Not less than 1 in 20 samples should be subject to replicate analysis. Replicates should conform to pre-determined performance characteristics. This procedure is indicative of errors associated with sub-sampling and analysis.

The rates of application quoted above are likely to be the minimum rates of usefulness of the internal quality control procedures.

4.5.4 *Recommended analytical methods*

Methods of analysis suitable for the more common site contaminants are listed in Table 4.14. None of the methods are specifically recommended by the publishing body as suitable for contaminated site investigation but they are recommended for use in matrices similar to those found on contaminated sites. Some of the methods have published performance criteria and that is indicated in the table, although in some cases the published criteria are very limited.

Users of those methods do therefore have some defined criteria against which to assess their own performance in operating the method. For those methods without criteria it is necessary that internal validation data is produced in assessing the suitability of the method for the individual application.

In considering the use of any of the methods the analyst needs to pay

Table 4.14

Analytical parameter	*Method reference*	*Matrix for which method is recommended by publishing body*	*Performance characteristics published*
Extractable or plant available cadmium, copper, lead, manganese, nickel, zinc	41, 53	Soils, sewage sludge treated soils	Yes (Ref. 53)
Fluoride	70	Waters, effluents sludges, soils, plants	Yes
Mercury	41, 71	Waters, effluents, Sludges, soil, plants	Yes (Ref. 72)
Sulphate	41, 72	Soil, waters, effluents and solids	Yes (Ref. 73)
Phenols	73, 74	Waters, effluents, Waste waters	Yes
'Total' cadmium, chromium, copper, lead, nickel, zinc	51, 41, 75	Soil, sewage sludges	Yes (Refs. 51, 76)
Pesticides/poly-chlorinated biphenyls	75	Sediments	Yes
Polycyclic aromatic hydrocarbons	75	Sediments	Yes
Sulphide	75	Sediments/solid wastes	No
Cyanide	74, 75	Waste waters	Yes
Ammonium	75	Sediments	Yes

particular attention to the difference in matrix between his application and that for the recommended use of the method.

4.6 Interpretation of analytical data

4.6.1 *General interpretation requirements*

The weakest link in the whole process of site investigation is likely to be the basis on which data is interpreted.

The DoE Guidance Note on Contaminated Land (1) poses six questions which must be answered by the assessment viz:

(i) What contaminants are present and in what concentration?
(ii) Which hazards are associated with those contaminants?
(iii) What is the intended use of the site?
(iv) Which hazards are likely to affect that use?
(v) What can be done to overcome them or minimize their importance?
(vi) Would a different intended use eliminate the hazards more effectively?

Such however is the lack of understanding of the quantitative effect of most contaminants on:

(a) The safety of site workers
(b) The integrity of building systems such as concrete and plastics
(c) The safety of future site occupants with respect to contact and ingestion
(d) The effect on plants
(e) Miscellaneous effects such as odour nuisance

that the investigation team will be required to exercise a great deal of value judgement in interpreting data in a reasonable and cautious manner.

Indeed site investigation has a vital role in helping to define the degree of caution needed to safely develop a site.

4.6.2 *'Background' and 'typical' levels*

To assist interpretation of site analytical data comparison with data from other sources (i.e. for naturally and widely occurring contaminants) and with guideline values can be valuable.

There are many sources of data for levels of metallic contamination existing in soils and rocks in the UK, some of which relate to 'background' levels (i.e. levels in areas well away from sources of contamination) and some to more 'typical' levels which may be found in many parts of the UK. Examples of such data sources are in References (54, 76, 77, 78).

Based partly upon literature indications and partly upon our own experience of soil metal level measurements, the author's laboratory uses the

Table 4.15

Metal (total)	*Guideline level mg/kg*
Lead (Pb)	500
Cadmium (Cd)	3
Copper (Cu)	200
Zinc (Zn)	300
Chromium (Cr)	500

following values of metal levels (Table 4.15) as likely to be found in many situations not associated with contaminated sites. Where metal levels on contaminated sites are found which are lower than these guideline levels then significant contamination is judged not to exist.

4.6.3 *Guideline levels*

In the UK values for assessing the concentrations of metallic and other contaminants have been suggested by the Department of the Environment, Greater London Council Scientific Branch, and Harwell. DoE tentative 'trigger concentrations' for inorganic contaminants are given in Reference 1.

Harwell recommended 'undesirable' and 'unacceptable' values for gas works site contaminants are given in Reference 9.

Figure 4.9 Otterspool riverside promenade, Liverpool, developed from a refuse disposal site. © City Engineer's Department, City of Liverpool.

Figure 4.10 A housing development on a former gasworks site

GLC typical values for 'uncontaminated' and 'contaminated' soils for a range of inorganic and organic contaminants are given in Reference 19.

These various guidelines can all assist the site investigation team in reaching their own reasonable and cautious value judgement. Clearly through, in the long term, greater central guidance from the DoE would assist in ensuring greater uniformity of approach to interpretation, and also encourage debate and examination of guidelines values and interpretation decisions.

4.6.4 *Mathematical methods of assessing data*

4.6.4.1 In view of the lack of any nationally accepted 'action' levels for interpreting contaminated site data, and the vagueness of DoE 'trigger' values, a cautious approach must be taken to assessment of site data.

However it should be part of the role of the site investigation team to improve the usefulness of data produced as well as indicate its uncertainties—this can only be of help in giving perspective to a situation for which no strict interpretation rules apply.

4.6.4.2 Initial examination of the overall data and of the sample duplicate data (section 4.3.7) may assist in regarding a contaminated area as producing one population of data or several populations to be considered individually.

For methods of evaluation the reader is referred to standard texts such as references 61, 79, 80.

Given the group of data to be evaluated the following techniques can be useful.

4.6.4.3 From the mean and standard deviation of the population of data can be calculated the two outer limits that contain nearly all of the population values. These are known as the statistical tolerance intervals (k). In most cases the site investigation would only be concerned with the maximum outer limit.

Suppose for example 50 measurements of lead are made on a site such that the following are estimated:

Mean	250 mg/kg
Standard deviation	125 mg/kg (i.e. coefficient of variation 50%)

From BS 2846: Pt. 3 (81) the maximum tolerance interval is given by:

$$x \pm ks$$

where x is the mean lead level
s is the calculated standard deviation
and k is a constant reflecting the confidence level to be chosen.

At the 95% probability level it can be calculated that 95% of the population is below the maximum tolerance interval of 500 mg/kg. If say, the DOE 'trigger value' for lead in gardens of 550 mg/kg were considered to be important in assessing the data, then there would be a reasonable degree of certainty that the trigger value would not be significantly exceeded.

4.6.4.4 The accuracy of individual results has been expressed in useful terms by various authors. Cheeseman and Beard (82) defined 'target accuracy'. McFarren *et al.* (83) defined 'total error' and Brown and Black (69) referred to 'uncertainty'. The three definitions have broadly similar characteristics and that of Brown and Black equates to a definition of uncertainty of an individual result as

$$U = (100 + B \pm 2C)$$

where B = bias = 100 − recovery
C = coefficient of variation calculated from analysis of duplicate samples

Suppose for example in the case of measurement of cadmium on a site recovery of the analytical method was estimated at 95% and coefficient of variation of duplicate samples at 15% then the uncertainty in an individual measurement would be 75% to 135%.

4.6.4.5 Clearly many other techniques are available to:

(i) Estimate sources of variance and pursue improvements in sub-sampling and analytical techniques

(ii) Estimate the overall uncertainty of data
(iii) Test the data against comparative values and others.

Consideration must be given to any technique which may improve the use of data, but in many cases it will be appropriate to take a broader and less detailed look at the investigation data. Plotting of maximum values, identification of trends etc. will be important but the extension of any data analysis must be seen in relation to the major weakness in site investigation—the lack of action levels for data interpretation. The maximum uncertainty lies in this area and cannot be quantified. Of all the facets of contaminated land study and reclamation this is the one in most urgent need of national attention.

References

1. Interdepartmental Committee on the Redevelopment of Contaminated Land (ICRCL) (1983) Guidance on the assessment and redevelopment of contaminated land. ICRCL 59/83. Department of the Environment.
2. JURUE (1984) Identification and assessment of contaminated land. Ecotec Research and Consulting Ltd., Birmingham.
3. Barry, D.L. (1982) Prospective surveys for contaminated land. WS Atkins Group, Epsom, Surrey.
4. Duffy, R. (1984) Survey of contaminated land in Wales, HMSO.
5. Dumbleton, M.J. and West, G. (1976) Preliminary sources of information for site investigation in Britain. Department of the Environment. TRRL Report LR403.
6. Dumbleton, M.J. (1979) Historical investigation of site use. Conference on the Reclamation of Contaminated Land. Society of Chemical Industry 1979.
7. Anon. (1983) Draft British Standard Code of Practice for the Identification and Investigation of Contaminated Land. Ref 83/55992, British Standards Institution, London.
8. Aldred, J.B. and Lord, D.W. (1984) *Chem. and Ind.* **17**, 624–627.
9. Wilson, D.C. and Stevens, C. (1981) Problems arising from the redevelopment of gas work and similar sites. AERE R 10366.
10. Pearce, K.W. and Finnecy, E.E. (1983) In: *Proc. Conf., Reclamation of former Iron and Steelworks Sites*, Windermere. Durham County Council.
11. Notes on the redevelopment of landfill sites. (1978) ICRCL 17/78, Department of the Environment.
12. Notes on the redevelopment of sewage works and farms. (1979) ICRCL 23/79, Department of the Environment.
13. Notes on the redevelopment of scrap yards and similar sites. (1980) ICRCL 42/80, Department of the Environment.
14. The Control of Pollution (Special Waste) Regulations (1980) SI 1709, HMSO.
15. Beckett, M.J. (1983) A systematic approach to the redevelopment of contaminated land. In: *Proc. Conf., Reclamation '83*, Industrial Seminars Ltd.
16. Davies, O.L. *Statistical Methods in Research and Production.* Oliver and Boyd, London.
17. Bell, R.M., Gildon A., and Parry G.D.R. Sampling strategy and data interpretation for site investigation of contaminated land. In: *Reclamation of Former Iron and Steelworks Site*, G.P. Doubleday (ed.), Durham County Council.
18. British Standard 5930: 1981 'Site Investigations'. British Standards Institution, London.
19. Kelly, R.T. (1980) Site Investigation and Materials Problems. In: *Reclamation of Contaminated Land*, SCI.
20. Waterhouse, P.S. (1980) The sampling and analysis of polluted sites. *Ibid.*
21. Lowe, G.W. (1980) GLC Development at Thamesmead. *Ibid.*

22. Carpenter, R.J., Goaman, H.F., Lowe G.F. and Pecksen, G.N. (1985) Guidelines for site investigation of contaminated land, London Environmental Supplement No. 12.
23. Smith, M.A. and Ellis, A.C. (1986) An investigation into methods used to assess gas works sites for reclamation. *Reclamation and Revegetation Research*, (In Press).
24. Keith, H.L., Crummett, W., Deegan J., Libby, R.A. Taylor, J.K. and Wentler, G. (1983) Principles of Environmental Analysis. *Anal. Chem.*, **55**, 2210–2218.
25. Methods for the examination of waters and associated materials (1980) General principles of sampling and accuracy of results, Department of the Environment/National Water Council, HMSO.
26. Mason, B.J. (1983) Preparation of soil sampling protocol techniques and strategies. EPA-600/4-83-020.
27. Barth, D.S. and Mason, B.J. (1984) Soil sampling quality assurance user's guide. Environmental Research Center, Las Vegas, N.V. EPA 600/4-84-043.
28. Kratochvil, B. and Taylor, B.K. (1980) Sampling for Chemical Analysis. *Anal. Chem.*, **53**, 944A–938A.
29. Youden, W.J. (1967) The role of statistics in regulatory work. *J.A.O.A.C.*, **50**, 1007–1013.
30. The Health and Safety at Work Act (1974), HMSO.
31. The Health and Safety (First Aid) Regulations (1981), HMSO.
32. British Standard 4275:1974 'Recommendations for the selection, use and maintenance of respiratory protective equipment'. British Standards Institution, London.
33. Health and Safety Executive Hazardous Substances Division 25, Chapel Street, London NW1 5DT.
34. Personal communication, County Analyst for Greater Manchester.
35. Dumbleton, M.J. and West, G. Guidance on planning, directing and reporting site investigations. TRRL Report LR625.
36. Thorburn, S., Colemutt, B.A. and Douglas, S.S. (1979) *Intern. Journ. Environ. Anal. Chem.*, **6**, 245–254.
37. McBean, E.A. and Farquhar, G.J. (1980) *Water Air Soil Pollution*, **13**, 157–172.
38. Griesinger, B. (1977) *ISWA Inf. Bull.*, **21**, 6–13.
39. Methods for the examination of waters and associated materials (1977) The sampling and initial preparation of sewage and waterworks' sludges, soils, sediments and plant materials prior to analysis, Department of the Environment/National Water Council, HMSO.
40. The Macaulay Institute for Soil Research, Aberdeen. (1971) Spectrochemical methods for the analysis of soils, plants and other agricultural material. Consultative Committee for Development of Spectrochemical Work, Bulletin No. 2.
41. Ministry of Agriculture, Fisheries and Food (1981) The analysis of agricultural materials. A manual of the analytical methods used by the Agricultural Development and Advisory Service, HMSO.
42. Frank, U., Gruenfeld, M., Losche, R. and Laformara, J. (1980) 'Mobile laboratory safety and analysis protocols used at abandoned chemical waste dump sites and oil and hazardous chemical spills' *Proc. Conf., Hazardous Materials Spills*, Louisville, Kentucky, 259–263.
43. Freestone, F., Gruenfeld, M. and Wilder, I. (1978) EPA's mobile laboratory and treatment system responds to hazardous spills. *Industrial Water Engineering*, 18–23.
44. Losche, R. and Urban, M. (1978) 'Development and use of a mobile chemical laboratory for hazardous material spill response activities' *Proc. Conf., Hazardous Materials Spills*, Miami Beach, Florida, 311–314.
45. *Directory of Consulting Practices*. (In Press.) Royal Society of Chemistry, Burlington House, Piccadilly. London, W1V OBN.
46. Montgomery, R.E., Remeta, D.P. and Gruenfeld, M. (1985) Rapid on-site methods of Chemical Analysis. In: *Contaminated Land, Reclamation and Treatment*, M.A. Smith. (ed.) Plenum, New York and London.
47. Thornton, S., Moorcroft, S. and Watt, J. (1980) Cadmium at Shipham: a unique example of environmental geochemistry and health. *Trace substances in Environmental Health*, **14**, 27–37.
48. *Catalogue of BCR Reference Materials* (1984) Commission of the European Communities. Brussels, Belgium.
49. British Standard 1881:1971 Methods of testing concrete Part 6. Analysis of hardened concrete, London, British Standards Institution.

50. Carlton–Smith, C.H. and Davis, R.D. (1983) An inter-laboratory comparison of metal determinations in sludge-treated soil. *Wat. Pollut. Control.*, 544–556.
51. Methods for the examination of waters and associated materials (1981) Cadmium, chromium, copper, lead, nickel and zinc in sewage sludges. Department of the Environment/National Water Council, HMSO.
52. Methods for the examination of waters and associated materials (1979) Atomic absorption spectrophotometry, an essay review. Department of the Environment/National Water Council, HMSO.
53. Methods for the examination of waters and associated materials (1983) Extractable metals in soils, sewage sludge treated soils and related materials. Department of the Environment/National Water Council, HMSO.
54. Ministry of Agriculture Fisheries and Foods (1980) Inorganic pollution and agriculture. *Reference Book*, 326, HMSO.
55. Report of the Working Party on the Disposal of Sewage Sludge to Land (1977) Department of the Environment, HMSO.
56. The Toys (Safety) Regulations (1974) and the Pencils and Graphic Instruments (Safety) Regulations (1974), HMSO.
57. Wilson, A.L. (1970) The Performance characteristics of analytical method I. *Talanta*, **17**, 21–29.
58. Wilson, A.L. (1970) The Performance characteristics of analytical methods II. *Talanta*, **17**, 31–44.
59. Wilson, A.L. (1973) The Reference characteristics of analytical methods III. *Talanta*, **20**, 725–732.
60. Wilson, A.L. (1974) 'The Performance characteristics of analytical methods IV.' *Talanta*, **21**, 1109–1120.
61. Caulcott, R. and Boddy, R. (1983) *Statistics for Analytical Chemistry*, Chapman and Hall.
62. Cheeseman, R.V. and Wilson, A.L. (1978) Water Research Centre Technical Report TR66, Water Research Centre, Stevenage, Herts.
63. US Environmental Protection Agency (1979) *Handbook for Analytical Quality Control in Water and Waste Water Laboratories*, EPA 600/4-79-019.
64. Garfield, F.M. (1984) Quality Assurance Principles for Analytical Laboratories. Association of Official Analytical Chemists Washington, USA.
65. Lord, D.W. (1984) Analytical quality assurance in United Kingdom Public Analysts' Laboratories. *Analyt. Proc.*, **21**, 351–412.
66. National Physical Laboratory (1983) Certified Reference Materials and Transfer Standards. Department of Industry, London.
67. US Environmental Protection Agency (1979) Methods for Chemical Analysis of Water and Waste. EPA 600/4-79-020.
68. Brown, K.W. and Black, S.L. (1983) Quality assurance and quality control data validation procedures used for Love Canal and Dallas Lead soil monitoring programs. *Environmental Monitoring and Assessment*, **3**, N2, 113–22.
69. Di Domenico A, Silvano, Viriano, Zapprori (1983) Ecotoxicological and environmental safety 4, **3**, 339–345.
70. Methods for the examination of waters and associated materials (1978) Fluoride in water, effluents, sludges, plants and soils. Department of the Environment/National Water Council, HMSO.
71. Methods for the examination of waters and associated materials (1978) Mercury in waters, effluents and sludges. Department of the Environment/National Water Council, HMSO.
72. Methods for the examination of waters and associated materials (1979) Sulphate in waters, effluents and solids. Department of the Environment/National Water Council, HMSO.
73. Methods for the examination of waters and associated materials (1981) Phenols in waters and effluents. Department of the Environment/National Water Council, HMSO.
74. Gottfried, G.J. (1982) Application and evaluation of analytical procedures for trace metals, total cyanide and phenolics. EPA 600/54-82-009.
75. US Environmental Protection Agency (1984) Characterisation of hazardous waste sites. A methods manual. Vol 3. Available laboratory analytical methods. EPA 600/4-84-038.
76. The Royal Society of Chemistry (1982) "*Environmental Chemistry*, Vol. 2". London.
77. *Wolfson Geochemical Atlas of England and Wales.*

78. JURUE (1983) A structured survey to land contamination in Walsall. Ecotec Research and Consulting Ltd., Birmingham.
79. Barnet, V. and Lewis, T. (1978) *Outliers in statistical data.*, Wiley.
80. Davies, O.L. (1978) *Design and analysis of industrial experiments.* Longmans.
81. British Standard 2846: Part 3 (1975) Guide to statistical interpretation of data. London, British Standards Institution.
82. Cheeseman, R.V. and Beard, M.J. (1983) Regional analytical quality control. Water Research Centre, Stevenage, Herts.
83. McFarron E.F., Liska R.J. and Parker J.H. (1970) *Anal. Chem.*, **42**, 358–365.

5 Available reclamation methods

M.A. SMITH

5.1 Introduction

Consideration of what reclamation or treatment methods may be adopted can only properly be done once a comprehensive site investigation has been carried out (see Chapter 4). Environmental pollution is usually viewed in terms of a source and of pathways to a target at risk. Once a risk is identified, the options available are to eliminate the source, remove the target, or to cut or restrict the pathways (Figures 5.1 and 5.2). All three options are available, singularly or together, for dealing with contaminated land.

The method selected will depend on:

(i) The objectives of the remedial action as defined taking into account the type and extent of contamination, the potential threat to identified targets, and the intended use of the land (as explained elsewhere, the objectives of the treatment will include considerations other than purely environmental ones and the objectives may conflict)
(ii) The technical solutions available to meet those objectives, and
(iii) Economic and social factors.

Except in rare cases, the aim will be to find a solution that is either permanent or which offers long-term protection. The most common approaches in the UK have been:

(i) To allocate land use so as to minimize risk to the most sensitive targets (the 'land use option')
(ii) To remove the polluted ground for deposition elsewhere and to replace it with clean fill material,
(iii) To isolate the contaminated ground from targets on the surface by imposition of a cover of clean fill material, and
(iv) In the case of methane and other gases, to install systems designed to control the movement of gas in the ground and to prevent its entry into buildings.

These approaches will not, however, always be sufficient to meet all the objectives of a reclamation scheme. Other treatment options are available, as described below, and may be essential on some sites. Proper decisions can only be made if the objectives of the treatment are carefully defined and a systematic approach adopted to consideration of the options. A failure to do so may result in adoption of measures, which whilst very suitable from one point of view, may compound problems from another.

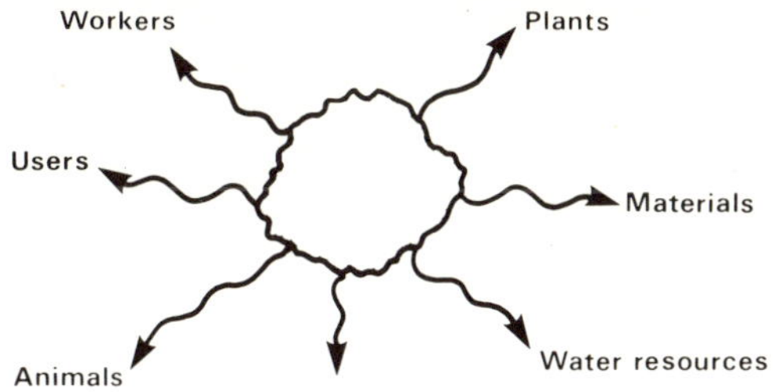

Figure 5.1 Contaminant source and potential targets

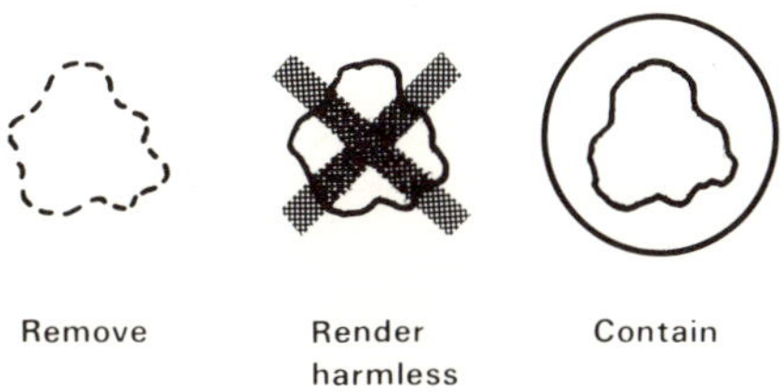

Figure 5.2 Basic options for dealing with contamination

Whilst the aim will usually be to achieve a permanent solution there may be times when a temporary or partial solution may be necessary in order to protect public safety or to bring a quick halt to ongoing pollution, e.g. of surface waters. Care must be taken to ensure that what was intended to be temporary does not become the final solution and that the temporary works do not make a permanent solution more difficult to achieve.

In practice, the primary objectives in the UK to date have usually been to protect man and buildings and to enable vegetation to be grown on the surface, although at times reclamation works have been undertaken specifically to reduce pollution of surface waters and consequential secondary pollution of agricultural land (1). A need to protect, or to clean up groundwater has not generally been seen as important although occasionally Regional Water Authorities have expressed concern about the potential for pollution of aquifers underlying contaminated sites as a result of reclamation or building works on them. About 25–30% of our domestic water supplies are, however, taken from ground sources and this, together with the implementation of EEC requirements to protect groundwater (2) may require that greater attention is paid to this in the future.

Whilst it will not usually be possible to give full consideration to the method of reclamation until after a proper site investigation has been carried out, it is important before beginning the investigation to consider data needs in relation to the possible options and to take account of any engineering or reclamation work that it has already been decided is essential.

It is essential in all reclamation activities to approach the task like any other engineering task: to define objectives, to quantify where possible, and to explicitly state any assumptions that are made.

The safety of workers and neighbours whilst work is in progress must be a paramount consideration (Chapter 8). Thought must also be given to the safety of those who may have to carry out site works in future years (Chapter 7).

5.2 The land use option

The targets at risk and the degree of exposure to the risk will depend upon the use to which land is to be put. For example inorganic pollutants such as lead and cadmium will present no immediate hazard to humans if covered by concrete (they will remain however to present a future problem if there is a change of land use in future years) but would be of major concern if present in garden soils. Such considerations may be taken into account on a macro-scale, i.e. one avoids zoning heavily polluted land for 'sensitive' uses such as housing (see also Chapter 9), or on a micro-scale within a site layout to try to ensure as far as possible that the more heavily polluted areas are beneath buildings, roads etc.

It is important that land use zoning is seen as a tool in dealing with contaminated land both to enable more satisfactory technical solutions to be adopted and to avoid undue cost in making a potentially unsuitable site fit for (say) housing. It must be recognized however that at times social and political factors may have to outweigh the purely technical and economic considerations.

A number of practical examples of the use of land use zoning to assist in dealing with contaminated land have been described by Smith (3). The cases cited include the Beaumont Leys development at Leicester (4), parts of the Greater London Council's Thamesmead development (5) and two small gasworks sites.

In order to make preliminary judgements on land use zoning it is essential for planners to have available information on the location of contaminated sites and this will only generally be available if site identification surveys are made, such as those carried out in Denmark, the Netherlands (see Chapter 1) and Wales (6) and carried out as pilots in other parts of the UK (7 and 8)—see also Chapter 9.

5.3 General considerations

5.3.1 *The 'typical' site*

Sites vary greatly in their size and the nature and extent of contamination. Beaumont Leys in Leicester covers 800 hectares and the polluted parts of

Thamesmead 250 hectares. However, for convenience and because of the limited space here consideration will be given to a 'typical' urban site covering no more than a few hectares at most and frequently less than a hectare in size. It will generally be in an urban setting and will typically be:

(i) An old pit or quarry filled with poorly consolidated refuse, rubble and undefined wastes, and often containing large amounts of combustible materials and permeated with methane, or
(ii) An old industrial site; the buildings will often have been demolished and the rubble, wastes and contaminants spread indiscriminately about the site in order to provide a level surface. There will be old foundations, drains and buried tanks, and areas of fill and made ground. The ground will be polluted to a considerable depth owing to spillages and deliberate dumping of wastes with a range of contaminants, which often include asbestos, lead, cadmium, etc., and mobile organic liquids such as oils, tars and solvents.

In considering how to deal with the site, the following are among the factors that will need to be taken into account:

(i) Present and intended topography and the relation of site levels to surrounding areas, roads, services etc.
(ii) Adjacent land areas (e.g. proximity of buildings)
(iii) Surface drainage, adjacent water courses, groundwater levels and movement, underlying aquifers
(iv) Propensity of the site for flooding etc.
(v) Location of existing services
(vi) Maximum depth of excavation required for services or foundations, (major services, especially sewers, usually have to be installed at considerable depths and this inevitably means digging into the contaminated material even if all the other works can be kept within any clean material that might be laid over the site)
(vii) The consequences of settlement within any imposed clean soil cover and of settlement in the underlying ground due to imposed loads from the cover or building
(viii) The safety of workers and neighbours during site works
(ix) Environmental impact of site works
(x) The significance of a future pollution incident on the site
(xi) Effect of building works (e.g. foundations and services) on any completed reclamation works, and
(xii) The significance of any future sites works (e.g. extensions to buildings, repairs to services or of any change of land use).

The last is likely to be an increasingly important consideration for whilst housing may be designed on the assumption that it will last for 100 years, some industrial buildings will have life times of 20 years or less before being demolished to make way for more up to date premises. If this requires major excavation and replacement of services then pollution buried safely beneath concrete or soil cover may re-emerge.

Indeed whilst the concept of burying immobile or relatively immobile contaminants beneath land development seems a generally acceptable and economic way to proceed, it is philosophically much more difficult to accept where the pollutants are mobile and free to move beneath the ground to affect adjacent land or nearby aquifers, or are capable of giving off volatile organic compounds.

Demolition of buildings on potentially contaminated sites frequently goes ahead without any regard to what is to follow. As a result contamination can be spread around the site and valuable information on plant layout can be lost. Demolition should be seen as part of the overall reclamation process to be undertaken in a controlled way in accordance with the relevant codes of practice (9) rather than as the last task for a plant manager and a few remaining workers charged with getting the maximum immediate value from the process.

On these urban sites contamination is unlikely to be the only problem. There are, as already pointed out likely to be engineering problems, and there may, in addition, be difficulties of access (important when materials have to be brought on and off-site) and a lack of basic infrastructure. It is important that the reclamation and redevelopment process is integrated to avoid unfortunate interactions because solutions to engineering problems may compound pollution problems, (e.g. ground consolidation may squeeze out gaseous and liquid pollutants or push putrescible matter below the water table).

5.3.2 *Defining the objectives of a reclamation scheme*

It is important that the objectives, both primary and the subordinate ones arising from them, are formally recognized and the options for meeting them systematically reviewed, preferably by a multi-disciplinary team.

A primary objective can usually be defined in terms of providing a development (e.g. housing, industry), a public amenity (e.g. public open space) or solving an environmental problem. In order to achieve this primary objective, a set of secondary and tertiary objectives will need to be defined and met. The secondary objectives (but in the end the most important) can be framed in terms of reducing the risk to identified targets to an 'acceptable' level. Subordinate, practical objectives, can then be derived taking into account such factors as those listed in the preceding section. The difficulties arise in properly identifying the targets, reconciling the interests of different parties to a reclamation scheme or development, in giving due consideration to longer-term as opposed to short-term objectives, and in ensuring that practical considerations such as ease of carrying out the work are taken into account.

The potential conflicts of objectives and/or interests can be illustrated by reference to a few examples.

(i) In the case of a housing development (public or private sector) the developer will aim to adopt remedial works at the lowest cost that will allow the designed development to go ahead. This may not be sufficient however for the householder (tenant or owner—with the recent right to buy legislation they now have the same interests) who has a longer term interest in being able to extend, add garages, introduce new fuel connections etc. and in being able to sell the house free of encumberances. It should be noted that it was assumed in some of the early public sector schemes that there would be permanent local authority control over any future actions on the site that might disturb the remedial works.

(ii) Similarly, for an industrial developer, works sufficient to enable a development to go ahead in the short term, such as the provision of a suitable depth of cover over the contaminants which remain in place, may simply put off the time when a more complete solution has to be carried out: it may be difficult to redevelop the site in 10 or 20 years time without carrying out further extensive remedial works. The ability of the owner or leaseholder to sell properties may be affected by such considerations.

(iii) A real example. The short term objective of reducing pollution of a water course was achieved by vegetation of the offending spoil heaps using a combination of top soil cover and toxin-resistant grasses. However, even apart from poor execution which left some areas of toxic material exposed, a permanent solution has not been achieved as the land cannot be safely grazed by animals (because the grasses contain elevated levels of metals and surface soils remain contaminated in places), nor can it be used as an amenity area. A longer-term approach would have been to either carry out sufficient work to achieve either of these objectives or to allow the land to be used, like much of that around it, for softwood forests.

(iv) It is worth noting that in the early days of derelict land reclamation the main objective when dealing with a difficult substrate (toxic or lacking in nutrients) was to turn it green at the lowest possible initial cost. The safety of those using the site, or the longevity of the reclamation work were not always given sufficient attention.

Costs will always be a factor but, as in many fields of construction, a minimization of initial capital costs can easily lead to increased long-term costs in future maintenance. Similarly a little more expenditure on the reclamation works can reduce costs in later stages of the development: for example if sufficient depth of cover is provided to allow excavation for the installation of minor services without injuring the effectiveness of the cover system, then the need to protect individual service runs will be obviated. Even if it costs a little more, the simplification of the construction process and reduced requirement for supervision at all times may well be worth paying for.

Similarly in reclamation of areas to be landscaped or used for gardens, a slightly higher initial cost can significantly reduce maintenance costs.

5.3.3 *The Building Regulations: The 'approved document' on contaminants*

A revised system of Building Regulations (10, 11) was introduced in England and Wales in 1985 under the Building Act 1984. Schedule 1 to the Regulations contains a set of 'Requirements' contained within various 'paragraphs'. In support of the Regulations a set of 'Approved Documents' have been produced. According to the introductory material to the Approved Document (12) dealing with contaminants:

> These documents have been approved by the Secretary of State as practical guidance to meeting the requirements of "Paragraphs in Schedule 1 to the Regulations", but there is no obligation to adopt any particular solution in the document if you prefer to meet the requirement in some other way.
>
> If a contravention of a requirement is alleged, then if you have followed the guidance then that will be evidence tending to show that you have complied with the Regulations. If you have not followed the guidance then that will be evidence tending to show that you have not complied. It will then be up to you to demonstrate by other means that you have satisfied the requirement.

The Approved Documents generally consist of

(i) A statement of the Requirement,
(ii) A statement of an 'Acceptable level of performance (including any essential definitions), and
(iii) Guidance on how this 'acceptable level of performance' may be achieved.

The Approved Document dealing with contaminants (11) contains, in addition, a useful Appendix (which is not part of the formal guidance) which describes the work of an expert adviser.

The wording of the relevant parts of the Approved Document on contaminants (12) is as follows:

> *Requirement:dangerous and offensive substances*
> C2, Precautions shall be taken to avoid danger to health caused by substances found on or in the ground to be covered by the building.
> *'Acceptable level of performance'*
> (it also covers C1, preparation of site and C3, sub-soil drainage):
> 0.1. To reduce the risks to their health and safety persons in buildings, and the stability of the buildings which they are in, should be safeguarded against the adverse effects of:
>
> (a) Vegetable matter, and
> (b) Groundwater, and
> (c) Contaminants in or on the ground to be covered by the building.
>
> 0.2. Contaminant means any material in or on the ground covered by the building (including faecal or animal matter) and any substance which is or could become toxic, corrosive, explosive, flammable or radioactive and so likely to be a danger to health and safety.

Whilst this innovation, and the guidance that goes with it, is a development to be welcomed, the requirement and the advice given has limitations.

Once a building has been erected, the occupants are only likely to be at risk from gaseous (e.g. methane) or other volatile substances. These may enter beneath a building or by some other route (e.g. service runs) from outside of 'the ground covered by the building' which is the subject of the requirement. The potential for mobility of the contaminants most likely to be a problem is not explicitly recognized in the Approved Document and the actions described as meeting the performance (section 2 of the Approved Document) may not provide adequate remedy.

Section 2 contains a Table listing various contaminants and the relevant action that should be taken of which there are basically four; none, removal, filling, and sealing, to be applied on their own or in combination. It is stated that 'the relevant actions will meet the performance. These actions assume that the ground will be covered by 100 mm of in-situ concrete laid over it.'

The actions are defined as follows:

> *Removal* means that the contaminant itself and any contaminated ground to be covered by the building should be taken out to a depth of 1 m (or less if the Local Authority agrees) below the level of the lowest floor....
> *Filling* means that the ground to be covered by the building is to be covered to a depth of 1 m (or less if the Local Authority agrees) with a material that will not react adversely with any contaminant remaining....
> *Inert filling* means the filling is wholly non-combustible and not easily changed by chemical reactions....

It is simplistic and potentially dangerous to suggest as Section 2 appears to do that all problems can be solved by either removing 1 m of contaminated material beneath the building or by covering the ground to a depth of 1 m, particularly without reference to the need to exclude mobile contaminants. The use of the Building Regulations to control the development of contaminated land is discussed in Chapter 9.

5.3.4 *Long-term effectiveness of remedial measures*

When the contaminants are not removed or destroyed, any remedial measure undertaken will be intended to prevent or impede the contaminants from reaching the targets at risk. These remedial measures must be designed to be durable and robust. Durable in the sense that they continue to perform in the desired manner for the length of time required, and robust in the sense that their effectiveness is not easily undermined by natural events such as flooding, subsidence and vegetation growth or unconscious intervention by man, for example excavation to install or repair services, or by bad workmanship at the time of installation. If the treatment is to be considered permanent, then it must be designed to last indefinitely, or at least for the lifetime of any building placed upon it. This is of course difficult to do at present as we do not have the experience on which to base our designs. There are only a limited number of sites more than a few years old reclaimed with full recognition of the contaminant problems, and none on which long-term monitoring or evaluation studies (see Chapter 7) have been carried out.

In considering the probable long-term effectiveness of a proposed remedial action the following should be taken into account:

(i) There will always be flaws in execution and the theoretical effectiveness will never be achieved
(ii) Environmental factors will generally serve to reduce effectiveness with time
(iii) Intervention by man, including change of land use in subsequent years, may reduce effectiveness, and
(iv) Contamination will not generally reduce with time (although methane is an exception here).

It is essential therefore that the time element is properly taken into account in the design of reclamation works and that, as with any building or construction work, the needs for maintenance and monitoring are also properly considered.

The question of long-term effectiveness, and the extent to which the various options discussed below are capable of providing long-term solutions to contamination problems, has been reviewed at length by Stief (13) and is discussed further in Chapter 7.

5.4 The options

The options for treatment can be classified in a number of ways. The basic question is whether the material is to be excavated or left in-situ. Once excavated the contaminated material may be:

(i) Deposited elsewhere
(ii) 'Cleaned-up' on-site or off-site, or
(iii) Treated to stabilize or 'fix' the contamination.

Treated soil ((ii) and (iii)) may be returned to the excavation, deposited elsewhere, or put to some beneficial use.

If left in place the options are:

(i) To do nothing other than to prevent access to the site and deal with any immediate environmental problem (e.g. remove any deposits of asbestos, control surface run-off)
(ii) Contain or isolate the site by superimposing cover and providing in-ground barriers to contaminant migration as necessary
(iii) To stabilize or fixate the contaminants in-situ, and
(iv) To clean up the soil in-situ.

Any of these treatment options where the contamination remains in place may require additional measures to be taken to control groundwater movement, and to contain or treat leachate and contaminated groundwater.

As already noted, the general approach to date in the UK has been to remove contaminated ground or to isolate it beneath a cover layer. However, as described below, extensive effort is being expended in other countries to develop other reclamation measures including the cleaning of soil to achieve a truely permanent solution.

The processes available for treatment (clean-up etc.) of soils are in principle the same whether carried out in-situ or after excavation. The options for treatment are therefore discussed further below under the headings:

(i) Excavation for deposition elsewhere
(ii) Soil treatment methods
(iii) In-situ treatment methods
(iv) Isolation or macro-encapsulation, and
(v) Control and treatment of groundwater and leachate.

This analysis follows closely the detailed and comprehensive review published (14) by the NATO Committee on Challenges of Modern Society Study Group on contaminated land and adopted for many of the papers to the Conference on Contaminated Soil held in Utrecht in 1985 (15). Advice on how to approach the selection of remedial measures in any particular case has been provided by Hoogendoorn and Rulkens (16). Lambert *et al.* (17) also provide a summary of all practically and conceptually available reclamation/detoxification methods and a commentary on their status and applicability.

5.5 Excavation

The first and most obvious solution will appear in many cases to be simply to excavate the contaminated material for deposition elsewhere and to replace it with clean imported fill as illustrated in Figure 5.3, or more properly in Figure 5.4, as the new fill must be placed using proper engineering techniques. In

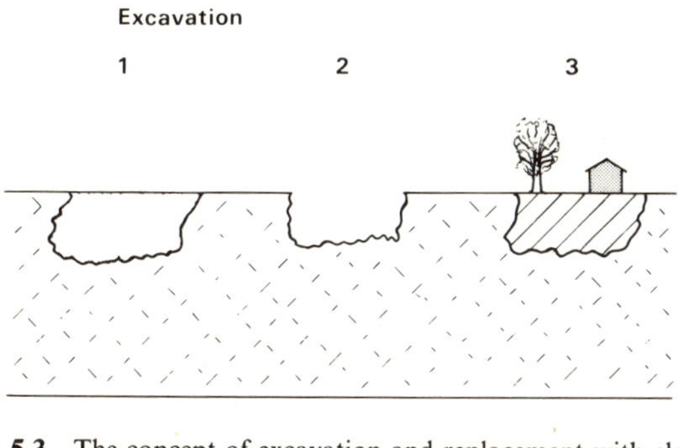

Figure 5.3 The concept of excavation and replacement with clean fill

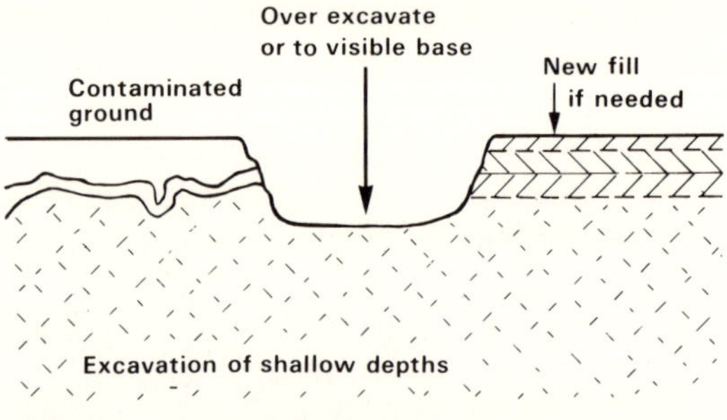

Figure 5.4 Excavation and emplacement of new fill

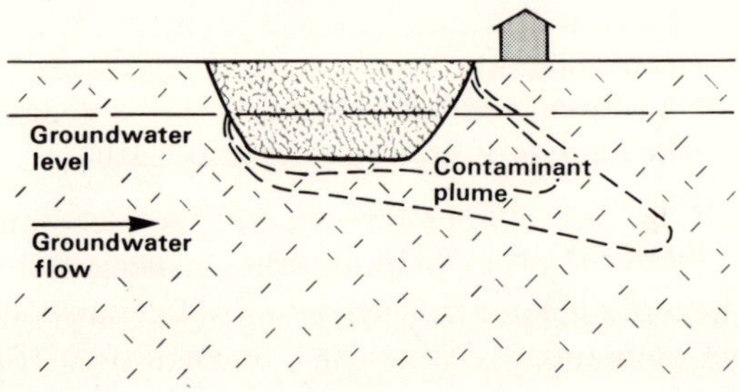

Figure 5.5 The complications that can arise from contaminant movement by groundwater

reality the process can be for more complex and a more realistic situation is illustrated in Figure 5.5.

Complications may arise because there may be:

(i) No sharp edge to the contamination
(ii) Contamination may have moved out of the site including beneath neighbouring buildings
(iii) Excavation may cause hydrogeological problems
(iv) Difficulty finding a disposal site
(v) Difficulty finding suitable clean fill material, and
(vi) Environmental problems—traffic movement, noise, atmospheric and surface water pollution, odour problems.

and because it may be necessary to

(i) Underpin adjacent buildings
(ii) Collect, treat and dispose of contaminated water
(iii) Temporarily lower groundwater levels
(iv) Protect rivers, streams etc.
(v) Monitor working environment and neighbourhood
(vi) Have chemical specialists on site
(vii) Take extensive health and safety measures to protect workers and neighbours, and

(viii) Provide additional remedial measures, e.g. to prevent contamination re-entering the site.

Notwithstanding these potential problems, excavation may well be the technically most satisfactory solution when the contamination is fairly shallow and the volume relatively small. Partial excavation may sometimes be necessary in order to either arrive at desired site levels or to make space for a clean cover layer. When the latter is done it is advisable to provide sufficient clean fill to enable minor services to be installed without digging into the contaminated soil. It is worth noting that when the contaminated soil is deposited elsewhere, it may in due course become a problem once again.

The problems which can be encountered are well illustrated by the case of the excavation of a former tar distillery site described by Armstrong (8). The approximately 2.6 hectare site, which was to be developed for housing, was contaminated extensively with phenolics, coal tars, sulphates and sulphides, and in parts with lead. Large tanks of tar and creosote were present and the groundwater was only about 0.3 m below surface level. The proposed reclamation scheme was as follows:

(i) Isolation of the site by complete enclosure with sheet piles, down to the base clay, at a depth of 3 m. This would inhibit the movement of contaminated groundwater on and off the site
(ii) The de-watering of the site by a pump and ring-main to a small scale treatment plant involving a sedimentation tank and three-stage petrol interceptor and then discharge to the foul main via a water authority trade effluent consent
(iii) Isolation of the contaminated zone by installation of a bunded clay wall down to base clay in the areas where the extent of contamination could not be clearly identified. Excavation would then continue up to the bund as far as was necessary
(iv) Excavation of all the material within the isolated area down to clean clay with disposal to a notifiable waste site (Pitsea) with an initial estimated volume of 40 000 m^3; and
(v) Backfill with clean clay.

The Environmental Health Department called a meeting involving the developer, statutory bodies and other parties with an interest, to ensure that all parties fully understood what was expected of them. A responsible person was identified in each of the statutory undertakers who could act in a liaison capacity when required. This group included representatives from the developer, the Local Authority, the Waste Disposal Authority, the Water Authority and the Health and Safety Executive Factory Inspectorate.

In practice it was necessary to remove a total of about 95 000 m^3 of material from the site, of which about 75 000 m^3 was designated as special waste within the terms of the Control of Pollution (special waste) Regulations 1980. Lorry movements were over 100 per day and special arrangements had to be made to assure the safety of children attending a nearby school.

It was considered that off-site contamination might arise from material falling from lorries or air-borne dust and odour.

The problem of material fall from lorries was dealt with in a number of ways:

(i) On-site arrangements were made to ensure lorries did not drive over contaminated material
(ii) A wheelspinning and hosing facility was installed at the site
(iii) Manual cleaning of material around the gate was carried out by the contractors
(iv) Daily mechanical cleaning of roads around the site was carried out by Local Authority sweepers, and
(v) All lorries were fully sheeted before leaving the site.

When certain areas of the site were worked, odours of phenolics and occasionally hydrogen sulphide were detectable in many of the local dwellings although on site levels as monitored by the site chemist remained well below the threshold limit values.

Removal work was carried out during the long hot summer and dust clouds caused by lorry movements were considerable. The Local Authority therefore undertook monitoring in three premises along the site boundary for coal tars, phenolics and lead, on a daily basis. Levels of atmospheric coal tars measured at the site boundary gave some cause for concern initially, and on a number of days were measured up around the threshold limit value of 0.2 mg/m^3. These results were somewhat deceptive in that the dust collected would not all be respirable and that the cyclohexane extraction procedure rather than the benzene method was adopted. Nonetheless, when reported to the Health and Safety Executive the factory inspector involved arranged for monitoring to be carried out by the contractors at the site boundary on a daily basis. As the weather situation deteriorated so a natural alleviation of the dust occurred.

Lead levels in the area were also raised to the upper end of what one might expect in an urban area and occasionally above the EC limit value of two μg/m^3.

5.6 Soil treatment after excavation

5.6.1 *Introduction*

The treatment of soil (or more generally material excavated from the ground) to remove or destroy contaminants, or to stablize or fixate the contaminants has hardly received attention in the UK, but the former is the preferred approach in the Netherlands and is receiving increasing attention in other

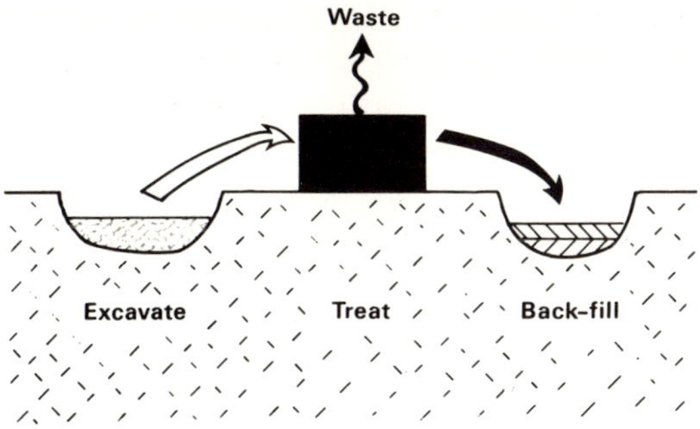

Figure 5.6 The concept of soil clean-up

countries. The 'clean-up' option is important because it offers an ultimate solution although there will often be an associated waste stream to dispose of. The processes can be carried out using mobile or temporarily erected plant on-site (Figure 5.6) or at a central treatment facility (the approach now favoured in the Netherlands). The treated soil may be returned to the excavation, disposed-of as a waste, or used as a fill or some other beneficial purpose. One of the unfortunate side effects of many of the treatment processes is that the soil is rendered inert and is not suitable for use as a growing media without amendment with fertilizers and organic matter.

It will be recognized that because the soil must be excavated before treatment and may have to be transported to the treatment plant, many of the problems associated with excavation described above may be encountered. In addition there will be the need to thoroughly evaluate the likely behaviour of contaminated soil in the treatment process chosen and to consider the environmental impact of the operating plants. A decision to adopt such a process will also raise the question of how clean the soil/treated material must be before it can be replaced in the ground or put to some other beneficial use, see Chapter 1.

Soil treatments can take a number of forms:

(i) Physical
(ii) Thermal
(iii) Chemical
(iv) Microbial
(v) Stabilization/solidification

While most of these involve processing the waste in some form of plant there are some simpler actions that can be classed as (on-site) soil treatment. For example, at the Beaumont Leys sewage farm at Leicester, before any construction of houses or factories occurs all top-soil is stripped and taken to stockpiles. In doing this the soil is separated into two streams on the basis of chemical analysis obtained in a comprehensive site investigation. Only that with an acceptably low degree of contamination is re-spread, for example on garden or allotment areas. Furthermore, disposal to stockpile and eventual recovery are carried out in such a way that the soil is mixed so that variations within the soil are reduced (4). Control has been exercised almost entirely in terms of cadmium content with limits being set of 3, 5 and 12 mg/kg for large gardens and allotments, small gardens, and public open space respectively (levels on the site range up to 60 mg/kg). Surplus soil with high cadmium contents has been used for landscaping public open spaces and is covered before grassing with soil within the 12 mg/kg limit. The whole process has been carefully monitored.

A classification of the conceptually available processes is given in Table 5.1. A comprehensive review of the theoretically available treatment methods has been provided by Rulkens *et al.* (19).

A variety of studies and practical examples of soil clean-up technologies are contained in the proceedings (15) of the conference on contaminated soil held in Utrecht in 1985. These include reviews of thermal treatment methods by Leer (20), microbial methods by Kreuk (21), and of extractive methods by Assink (22). The latter includes reference to clean-up of soils contaminated with gasworks residues (cyanides and polynuclear aromatic compounds).

Most of the processes now operating in the Netherlands and elsewhere are based on the application of technologies previously developed for mineral processing where an ore containing very low concentrations of a valuable

Table 5.1 Soil treatment options

A. Physical
- Solvent leaching
- Gravity separation
- Particle sizing
- Settling velocity
- Magnetic
- Flotation

B. Chemical
- Neutralization
- Chemical oxidation
- Chemical reduction
- Hydrolysis
- Electrolysis
- Ozonation
- Photolysis

C. Thermal
- Direct heating
- Indirect heating
- Incineration
- Steam stripping

D. Microbial

E. Stabilization/Solidification
- Cement-based systems
- Polymer-based systems

metal is treated to obtain a sufficiently concentrated feedstock for smelting etc. or treatment of wastes or contaminated water. In some cases equipment designed for other purposes, e.g. preparation of tarmacadam, or manufacture of lightweight aggregates (25) has been adopted to the new task. There are acknowledged limitations with many of the processes: typically sandy soils are easier to treat by physical means than are clayey or organic (peaty) soils and they are more easily applied to soils with a single contaminant than to those with a range of contaminants, some still leave a significant volume of (more) contaminated material for further treatment or disposal and one of the unfortunate side effects of many of the treatment processes is that the soil is rendered inert and is not suitable for use as soil without amendment with fertilizers and organic matter. Nevertheless, the processes have reached a relatively advanced stage in a very short time (about 3 years since the Dutch programme started).

5.6.2 *Extraction and separation techniques*

In solvent extraction the contaminated soil is mixed with an extracting agent (in general an aqueous solution but sometimes an organic solvent). Potential applications include the removal of metals such as cadmium, copper, zinc, nickel, chromium, arsenic, antimony and lead using a mineral acid solution and lead, zinc and organo-metallic compounds, and some cyanides using sodium hydroxide solution. Hydrocarbons and halogenated hydrocarbons can also be removed (19).

Contamination is often preferentially present in the finer or coarser fraction of the soil or the organic components (e.g. humus). Contamination can therefore be removed in some cases using a process which separates the soil into fractions on the basis of specific gravity, or of particle size, or settling velocity.

The review by Assink (22) includes an assessment of the applicability of various methods to particular forms of contamination, practical limitations, and factors to be considered when selecting what method, if any to adopt. Costs are estimated to be within the range 80 to 300 Dfl/tonne (about £20 to £75/tonne) exclusive of excavation and transport costs. Assink describes a number of operational installations used for the treatment of soils contaminated with cyanides and polynuclear aromatics (gasworks residues), zinc, lead, oil, and chlorinated hydrocarbons. A number of those described are containerized, allowing for re-siting if necessary. Capacities are of the order of 15–20 tonnes/hr.

The UK is not without some experience of related processes. The secondary recovery of coal from colliery spoil tips helps in the reclamation of such sites. A UK firm (23) has developed a mobile plant for the recovery of metal values from metalliferous mine wastes. It has been used to obtain a commercial return from the wastes but not yet as a tool in a site reclamation process where on its own it would not be economic. After processing the waste is still highly contaminated and would require safe disposal but the money gained from the metal values would off-set the cost of the overall reclamation scheme.

5.6.3 *Thermal methods*

There are two principal ways of using heat treatment to remove contaminants:

(i) Removal of the contaminants by evaporation—either by direct heat transfer (convection or radiation) from heated air (or other gas), or an open flame, or by indirect heat transfer, and
(ii) Destruction of the contaminants by direct or indirect heating of the soil to an appropriate temperature.

The gas leaving the heating appliance must be treated, as appropriate, to destroy or remove any contaminants or unwanted products of combustion. A related process is steam-stripping in which steam is injected into soil to aid evaporation of relatively volatile contaminants which may be water soluble or insoluble.

According to Leer (20) four processes of thermal decontamination have been developed in the Netherlands. Three make use of rotary kilns with different combinations of direct and indirect heating, followed by incineration of the volatilized contaminants. The fourth uses a fluidized bed operating at 800–900 °C. Three of the processes have been shown to be capable of decontaminating polluted soils from gasworks (polycyclic aromatic hydrocarbons, complex cyanides), while all processes are capable of treating soils contaminated with hydrocarbons. Treatment of soils contaminated with chlorinated hydrocarbons presents more difficulty because of the possible formation of toxic chlorinated dibenzodins, and this is still under investigation. Capacities range up to 50 tonnes/hour for one of the processes with a sandy soil feedstock.

In the United States the US Environmental Protection Agency has developed a sophisticated mobile incineration system (24). This was intended, at least initially, for the treatment of PCB containing media including soils, and is consequently fitted with elaborate air pollution control systems. Less sophisticated is the system described by Hazaga *et al.* (25) in which an asphalt plant was adopted to treat soils contaminated with trichloroethane, trichloroethylene, toluene, ethyl-benzene and xylenes. The costs excluding excavation and backfilling were about US $116/short ton (about £90/tonne) compared to $126/short ton (about £100/tonne) for landfilling. Noland and Sisk (26) have described trials by the US Army of the incineration of soils contaminated with explosives using a 'transportable' (i.e. container based) rotary kiln.

No specific thermal processes for treating contaminated soil have been developed in the UK but waste oils and domestic refuse have been used to provide some of the fuel needs for cement manufacture. It is possible that this approach could be developed further to deal with some forms of contaminated soil.

5.6.4 *Chemical treatment methods*

As shown in Table 5.1 a whole range of chemical processes may theoretically be applied to the clean up of soils. For example cyanide may be oxidized to CNO^- or hydrolized to CO_2 and NH_3 or N_2, heavy metals may be oxidized or reduced to increase solubility (e.g. Cr^{3+} to CR^{6+}) or to enhance precipitation (Cr^{6+} to Cr^{3+}), and halogenated hydrocarbons may be hydrolyzed, oxidized or subject to electrolysis.

Two basic methods of treatment are possible,

(i) Treatment of the soil in suspension in a suitable liquid, and
(ii) Without slurrying.

General characteristics of such processes are that intimate contact between soil and chemicals is essential, contact times are frequently long, and treatment chemicals (which may themselves be contaminating) must be applied in excess to ensure complete detoxification.

The treatment employed to deal with the Ravenfield tips in Yorkshire (27) was a crude form of combined chemical treatment and stabilization. Acid tars were excavated from old quarries with a drag line, mixed with quick lime and replaced in the ground using normal civil engineering techniques. This land is now used as a recreation ground.

5.6.5 *Microbial treatment methods*

Microbial treatment appears to be among the most promising in the medium term for dealing with a whole range of organic contaminants including phenols, polychlorinated hydrocarbons, chlorinated hydrocarbons, oil and oil products, and dioxins (they may even have some application for treatment of contamination with certain metals). Biological treatment has been practised for decades in the composting of refuse, and treatment of sewage. There are two different ways of approaching the problem:

(i) As in the Blackburn case cited below, a community of microbes already existing on the site (and therefore presumably resistant to the conditions on the site) is collected and cultured in the laboratory, or
(ii) Strains of microbes are developed ('engineered') in the laboratory that are capable of metabolizing particular chemicals and/or tolerating particularly difficult conditions (e.g. high concentrations of compounds usually considered toxic).

Excavation of the soil prior to treatment, as in the Blackburn case, offers the greatest scope for creating optimum conditions and reasonable treatment times (perhaps a few months) should be obtainable in this way. The excavated soil can be placed in thin layers to various depths (treatment beds and windrows) using standard earth-moving techniques and microbes and nutrients applied using standard agricultural techniques such as fertilizing, ploughing and harrowing. The scientific basis of microbial treatment and some of the physical, engineering and other principles involved and reviewed in a series of papers (28) to a Seminar held in Birmingham in 1985.

Various other cases of research studies in progress are mentioned in a review of microbial decontamination of excavated soil in the Netherlands by Kreuk (21), and other cases and studies have been described by Mokveld *et al.* (29), Bewley (30) and Fredricks (31).

Among the contaminants which are biodegradable are (21):

(i) Straight-chain alkanes of up to C_{30} (branched-chain alkanes are degraded more slowly)
(ii) Chlorinated hydrocarbons
(iii) Aromatics such as benzene, toluene, xylene, phenols, and polynuclear aromatics (PNAs) with two or three rings (higher PNAs are degraded only slowly if at all), and
(iv) Some chlorinated aromatics.

Various research studies are in progress to widen this range.

In the UK one firm (32) claims to be able to clean suitable sites by microbial techniques. Microbes taken from the site are cultured in the laboratory and then mixed together with nutrients into the excavated soil. The first trial began in 1985 on a former gasworks site in Blackburn.

5.6.6 *Stabilization/solidification processes*

Stabilization and solidification processes do not, of course, remove the contaminants from the soils but rather are intended to physically or chemically eliminate or at least reduce, the hazardous nature of the soil (waste, sludge or liquid) so that it can be safely handled, stored or disposed of in some way, e.g. to land. The reduction in mobility or leachability of contaminants can be achieved by:

(i) A chemical reaction to form practically insoluble compounds
(ii) Isolating contaminants from water (on a micro-scale) by adding or forming hydrophobic compounds
(iii) Adding chemicals to fixate water and thus reduce micro-leachability
(iv) Vitrification, and
(v) Control of pH to minimize solubility (a special case of (i)).

Mechanism (iii) is the principle of most of the commercially available processes. Processes exist based on Portland cement, lime, pozzolanas (e.g. 'fly ash'), hydraulic slags, thermoplastics and a variety of other cementing or solidifying systems.

The main limitation of such processes is that the contaminants are not removed and the treated soil may remain a source of hazard. There are important doubts (33, 34) about the long-term performance of some of the available processes arising either from the basic physical and chemical characteristics or from the poor quality of application observed in practice. For example, the US Environmental Protection Agency found (33) that most heavy metals in five types of sludge failed to become, or hardly became, insoluble after treatment by some commercially available treatment processes. Each proposed use of such a process must clearly be critically examined.

5.7 In-situ treatment

In-situ treatment, in which the contaminated ground is treated to remove, destroy or fixate the contaminants without excavation, is conceptually the most attractive form of treatment since it offers an ultimate solution with minimal disturbance of the ground. On the other hand, it is at present the most difficult to achieve in practice, except for some relatively simple treatments at or near the surface. These latter include application of lime to raise the pH of soils containing elements such as cadmium to reduce uptake by plants (however it should be noted that because lime is leached from the soil regular subsequent treatments will be needed (35)), and deep ploughing to invert (say) the top metre of soil. Leaching with water or chemical solutions (a form of solution mining) has also been successfully applied (36, 37). A detailed review of in-situ treatment has been provided by Sanning (38) and Michelsen (39) lists a number of practical examples and theoretical possibilities.

It is possible (see Table 5.2) to envisage many of the processes considered for application to soil after excavation being applied in situ. There are a number of difficulties (38):

(i) Ensuring contact between treatment agent (including nutrients for microbial treatment) and soil,
(ii) Ensuring that treatment has been effectively applied, and
(iii) The existence of a waste stream that must be disposed of or treated.

In-situ treatments can be divided into:

(a) Those applied at or near the surface, and
(b) Those requiring injection of an agent at depth.

In-situ treatment of contaminated groundwater (i.e. groundwater downstream of the source that has become contaminated) is also possible and has been carried out in a number of cases (see for example refs. 39 and 40). These processes will not, however, be discussed here. Here we are concerned with treating the bulk of contaminated material in the ground which may include solids, and aqueous and non-aqueous liquid phases.

Reference has already been made to some forms of surface applied treatments. At greater depths agents (chemical, microbial or solidifying) have to be injected into the ground. The techniques to be employed are similar to those used in grouting and there are obvious potential problems in addition to that already mentioned of ensuring proper contact between agent and contaminants, in particular that pressure injection cannot usually be applied at depths of less than about 2 m without danger of fracturing and ground disturbance. The opportunities and problems have been reviewed in detail by Sanning (38) and by Barry (41).

The chemical processes which may be applicable (Table 5.2) are similar to those employable after excavation. If microbes are to be used then nutrients must also be provided. Microbial treatment seems likely to be promising in the medium term especially for treatment of superficial contamination at or near the surface (e.g. 0.5 m).

The most obvious agents to use for solidification are cementitious or chemical grouts. Compatibility between grouts and contaminants will clearly be a major consideration (42). Solidification may be desirable for reasons other than treatment of contamination, for example:

(i) To improve engineering properties of ground
(ii) To stabilize sludge etc. that presents a physical hazard, and
(iii) To prevent or stop combustion.

Since the agents that are applied may themselves be potential pollutants, in-situ treatment may only be possible if steps are taken to prevent migration of contaminant and treatment agent from the site, i.e. by creating a reaction vessel within the ground using in-ground cut-off barriers.

In view of the limited development of these techniques to date they will not

Table 5.2 In-situ treatment options

A. Physical
- Deep ploughing
- Leaching/chemical mixing
- Solidification

B. Chemical
- Neutralization
- Precipitation
- Oxidation
- Reduction
- Fixation
- Ion exchange

C. Other
- Heat treatment to fuse soil (in-situ nitrification—see (37, 43)
- Electrokinetic techniques see (37)

be reviewed further here. However, the successful application of in-situ techniques to deal with some soil pollution problems and with groundwater contamination should be kept in mind.

5.8 Macro-encapsulation/isolation

5.8.1 *Introduction*

Macro-encapsulation means putting a box around the contamination as shown in Figure 5.7. This may surround the whole site—top, sides and bottom, but frequently only partial encapsulation or containment is carried out. Sometimes other measures, such as control of the groundwater regime are an essential adjunct. The most frequently used component, at least in the UK, is a cover of some form.

It must be accepted at the outset that no containment system can be fully effective and that the effectiveness of containments is likely to fall with time. All barrier systems, like all dams, leak to some finite extent—there is a theoretical permeability, albeit very low, and likely flaws in installation. There may also be problems due to incompatibility of the materials used and the contaminants. For example a material such as clay with a very low permeability for water may be more permeable to other substances. In addition chemical reactions between clay and contaminant may increase permeability to water (44–46), and there may be chemical attack on the barrier material. The techniques available to construct in ground horizontal and vertical barriers have been reviewed by Childs (47) together with,

(i) The manner in which they may be employed with associated hydraulic measures (e.g. well-pumping systems) to contain pollution within a site or limit movement after it has left the site, and
(ii) Methods of treating contaminated groundwater extracted as part of such a combined scheme.

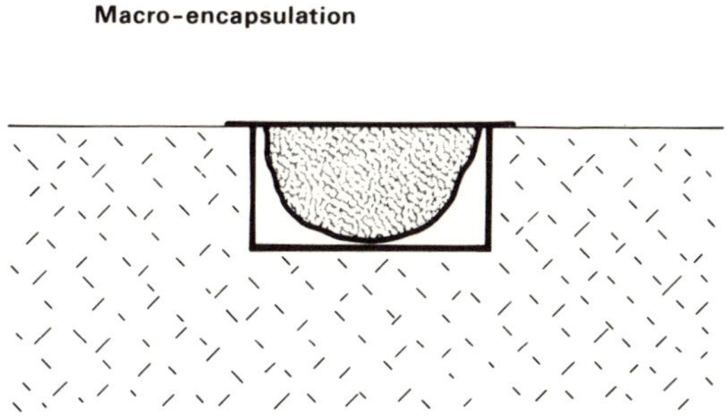

Figure 5.7 The concept of macro-encapsulation/containment

It is important before considering any in-ground barrier or containment system to consider very carefully the likely effects on the groundwater regime hence adequate hydrogeological data must be gathered. Computer simulations or models can be of great help in designing such systems (47).

5.8.2 *Vertical barrier systems*

Possible methods for forming vertical barriers are listed in Table 5.3. It is convenient to divide vertical barrier systems into:

(i) Those in which the ground is left relatively undisturbed, e.g. driven sheet piling or injected grout curtains (categories A and C in Table 5.3), and
(ii) Those in which material is excavated or otherwise removed from the ground. This extracted material may be contaminated. Feasible depths of application range from about 5 m for membranes (continuous or with a narrow trench) to 70 m for bentonite-cement 'deep-wall'.

The construction of vertical barriers using such techniques as grout curtains, concrete or metal diaphragm walls, and slurry trenches is a well established engineering practice, but, as already indicated above, polluted sites can present greater difficulties for a variety of reasons. A number of papers describing experimental studies and the application of barriers in practice have been described (44–46, 48, 49, 51). The US Environmental Protection Agency has published a manual on construction of slurry walls (52). Special methods for measuring the very low permeabilities required of

Table 5.3 Types of vertical barrier

A. *Driven and related systems*
Steel sheet piling
Vibrated beam slurry wall
Panel wall
Membrane wall
B. *Trench and other excavated systems*
Clay barrier
Natural material barrier
Bentonite-cement 'deep-wall' or barrier
Trenches and slurry walls with membrane
Narrow trench with membrane
Cast in-situ pile wall
Precast bentonite-concrete wall
Cement asphalt emulsions (45)
Systems based on 'chemically resistant' bentonites e.g. (45)
C. *Other systems*
Chemical grout curtains
Ground freezing (56)
Jet-grouting

Note: Childs (47) provides a summary description of each of the above methods (his categorization is however different).

barrier materials have been developed (53, 54) as have specially chemical resistant slurry media. Some novel methods of isolating sites have been attempted (55).

The choice of barrier systems will be governed by such factors as:

(i) Degree of integrity required
(ii) Depth required
(iii) How the junction is to be formed with an underlying low-permeability stratum or inserted horizontal barrier where this forms part of the containment
(iv) Resistance to corrosion and other forms of interaction with contaminants
(v) Geology and hydrogeology
(vi) Costs
(vii) Ease of installation on a particular site.

A pre-requisite of desigining a containment system is adequate knowledge of the hydrogeological characteristics of the site including data on the permeability of the various strata, their adsorption characteristics for contaminants and the ground flow pattern.

Beine and Geil (46) have discussed the processes involved in the passage of contaminants through, for example, a bentonite-cement slurry and the durability of such systems. Contaminant transport may occur by:

(i) Convection (i.e. movement of the fluid phase),
(ii) Dispersion (caused by deviations from the mean seepage velocity),
(iii) Molecular diffusion, and
(iv) Chemical and biological reaction between the contaminants and barrier material.

At the permeabilities achievable initially with bentonite–cement mixtures (10^{-9}m/s) seepage will occur primarily due to convection. Whilst some chemical mixtures may actually lead to a reduction in permeability, others will increase permeability. Systems based on Portland cement will be vulnerable to attack by sulphates and acids and the hydration reactions may be detrimentally affected by various organic substances, (e.g. sugars, phenols) and alkali-soluble metals (e.g. zinc, lead, arsenic).

5.8.3 *Horizontal in-ground barriers*

In some circumstances a naturally low-permeability layer beneath the site will be sufficient to provide a horizontal seal. Otherwise, sealing probably using some form of grouting technique is likely to be required. Methods identified (47) as of possible use in forming horizontal barriers are:

(i) Jet-grouting
(ii) Chemical grouting
(iii) Claquage grouting.

Installation of such below-ground barriers is clearly like to be difficult. When they are a critical part of the scheme it may be necessary to install two at different depths with monitoring points between to detect any failure of the upper most.

5.8.4 *Controlling infiltration*

A low permeability cap is likely to be required as part of the covering system that completes the scheme in order to minimize infiltration into the container that has been formed. Otherwise, if the rate of infiltration exceeds the leakage out, the 'container' may fill like a bath-tub and overflow. It may be necessary to install a groundwater pumping system to keep groundwater levels in the box low—the water table should preferably be lower than that outside so that any flow of groundwater occurs inwards. Any water pumped from within the closure is likely to require treatment or disposal.

5.9 Covering systems

The most common remedial action undertaken on sites in the UK is the superimposition of some form of cover (Figure 5.8). Unfortunately in very many cases very little thought has gone into their design. As explained in detail in Chapter 6 and elsewhere (56)—it is not just a case of putting half-a-metre-or-so of clay on the site. The functions that a cover must perform are demanding and often conflicting in their solution. A covering system must be designed and carefully installed if it is to function properly.

The primary function of the cover is to protect the potential targets (Figure 5.8) but they will often have other functions such as:

(i) To sustain vegetation, or
(ii) To fulfil an engineering role.

In order to do this the system as a whole or component parts of it will have

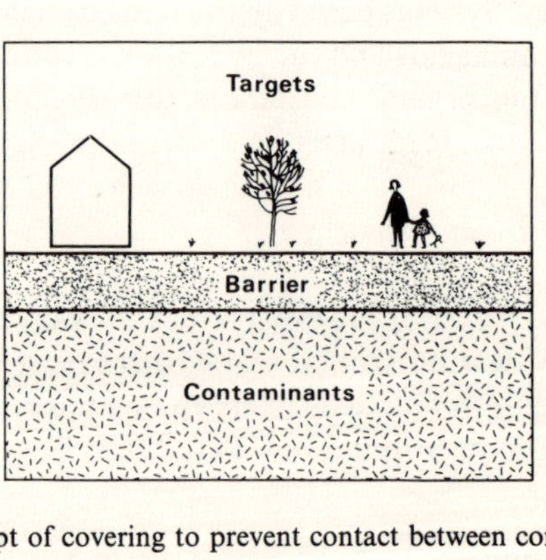

Figure 5.8 The concept of covering to prevent contact between contaminants and targets

to be able to do a number of things depending upon the particular case:

(i) To control gas movement, leachate, soil fluid movement, water ingress, erosion, slope stability, vermin, fires, odours, and
(ii) To minimize fire hazard, prevent dust blow, improve appearance, support vegetation, inhibit root penetration, prevent upward moisture movement, and to improve engineering properties.

The effectiveness with which they perform these various functions will be affected by many factors including:

(i) Ability to control water entry
(ii) Ability to prevent upward and lateral movement of contaminants
(iii) Ability to bind contaminants
(iv) Interaction with contaminants and biota
(v) Engineering behaviour.

Cover systems thus have to be designed with care and only after a through consideration of the physical and chemical properties of the soils and other cover materials that are available.

5.10 Execution of reclamation works

Successful completion of building and construction projects requires careful attention to design, specification and supervision whilst the work is being carried out, if the design objectives are to be met.

It is important that all involved in the reclamation process are full informed about the conditions on the site and about the purpose and function of the work to be carried out. First, because safety must always be the paramount consideration, and secondly because without this knowledge a contractor might inadvertently undermine the whole purpose of an expensively designed scheme either through poor execution of the main works or whilst carrying out some peripheral activity.

Two examples,

(a) A consultant says that half a metre cover should be provided over the whole site, the quantity surveyor calculates the volume required, and the contractor duly places it on site. Unfortunately, it is unevenly spread and he runs out before the site is finished. Toxic waste is left exposed. The result of poor specification and poor supervision.
(b) The contractor exports surplus clean material from one part of the site (e.g. clean soil or demolition rubble) at a profit. Later finding himself short of material he imports the cheapest available material which turns out to be contaminated.

As already pointed out above, the measures adopted must be capable of execution under practical conditions (i.e. take account of working tolerances in measurement and normal levels of workmanship).

5.11 Monitoring and evaluation of performance

The 'monitoring' of performance of civil engineering works and buildings is an accepted practice. At its simplest it is a programmed visual inspection to see where routine maintenance, for example painting, is required. At the more complex end it may involve the permanent instrumentation of a structure with an automatic system to indicate when any parameter goes beyond some pre-set limit. The need for such monitoring is not seen as undermining confidence in the design but rather as an essential component of the development helping to ensure its longevity and to ensure public confidence. Many reclamation schemes would benefit from such monitoring together with provision for programmed maintenance where appropriate.

Similarly there is much to be gained from a careful evaluation of the performance and behaviour of reclamation schemes over a prolonged period of time (there is a distinction to be made between monitoring which is designed to detect faults or trends leading to faults and an evaluation of performance which is a scientific study and likely to require a more sophisticated approach and greater expenditure of time and money). Such studies of real life structures have made significant contributions to the sciences of geotechnics and structural engineering.

5.12 Guidance available

Very little practical guidance of a detailed nature has so far been published in the UK.

The manuals of Liverpool University's Environmental Advisory Unit on reclaiming disused quarries (58), old metal mining sites (59), and urban sites (60) do contain valuable information, and the US Environmental Protection Agency has published comprehensive manuals (51, 60 and 61) on reclaiming contaminated and hazardous waste sites.

The sole other sources of useful information are the proceedings of specialist conferences see (15, 50 and 63).

References

1. Johnson, M.S. and Bradshaw, A.D. (1977) Prevention of heavy metal pollution from mine wastes by vegetative stabilisation. *Trans. Inst. Mining and Metallurgy*, (Section A), **86**, A47–A56.
2. EEC Directive on drinking water. *CEC Official Journal.*
3. Smith, M.A. (1982) Contamination and the built environment. *Proc. Seminar Policy Analysis*

for Housing and Planning, Planning and Transport Research Centre, Summer Meeting, Warwick, (PTRC, London 1982), 37–50.

4. Heeps, K.D. (1982) The reclamation of a disused sewage works. *Public Health Eng.*, **10** (4) 213–4 & 218.
5. Lowe, G.W. (1984) Investigation of land at Thamesmead and assessment of remedial measures to bring contaminated sites into beneficial use. Ref: 50, 560–564.
6. *Survey of Contaminated Land in Wales* (1984) Welsh Office, Cardiff.
7. Haines, R. (1984) Identification and assessment of contaminated land. Ecotec Research and Consultancy, Birmingham.
8. Barry, D. (1981) Prospective Survey of contaminated Sites. Atkins Research and Development, Epsom.
9. British Standard 6187: (1982) Code of Practice for demolition, British Standards Institution, London.
10. *The Building Regulations, 1985.*
11. *Building Act, 1984.*
12. *The Building Regulations, 1985*: Site Preparation and Resistance to Moisture: Approved Documents: C1/2/3 Site Preparation and Contaminants, C4 Resistance to Weather and Ground Moisture. HMSO, London.
13. Stief, K. (1985) Long-term effectiveness of remedial measures. In Ref. 14, 13–36.
14. Smith, M.A. (ed.), (1985) *Contaminated Land, Reclamation and Treatment*, Plenum, New York and London.
15. Assink, J.W. and Van den Brink, W.J. (eds.), (1986) *Proc. 1st. International Conf., Contaminated Soil.* Utrecht, 1985. Martinus Nijhoff, Dordrecht, 1–11.
16. Hoogendoorn, D. and Rulkens, W.H. (1986) Selecting the approptiate remedial alternative: a systematic approach. In: Ref. 15, 1–11.
17. Lambert, W.P., Bone, L.J. and Sisk, W.E. (1984) Evaluation of advanced technologies for treatment of contaminated soils. In: Ref. 50, 412–419.
18. Armstrong, G.D. (1984) Local authority involvement in the decontamination of a tar distillery site. In: *Contaminated Land: the London Experience*. London Environmental Supplement No. 7, Greater London Council, London, 10–13.
19. Rulkens, W.H. Assink, J.W. and Th. Gemert, W.J. On-site Processing of Contaminated Soil. In: Ref. 14, 37–90.
20. de Leer, E.W.B. (1986) Thermal methods developed in the Netherlands for the cleaning of contaminated soil. In: Ref. 15, 645–654.
21. de Kreuk, J.F. (1986) The microbiological decontamination of excavated soil. In: Ref. 15, 669–677.
22. Assink, J.W. (1986) Extractive methods for soil decontamination; a general survey and review of operational treatment installations. In: Ref. 15, 655–668.
23. Robertsons Research International, Llandudno, Wales.
24. Brugger, J.E. *et al.*, The EPA-ORD mobile incineration system: Present status' report No. 167, Oil and Hazardous Materials Spills Branch, US. Environmental Protection Agency, Edison, New Jersey.
25. Hazaga, D., Fields, S. and Clemons, G.P. (1984) Thermal treatment of solvent contaminated soils. In: Ref. 50, 404–406.
26. Noland, J.W. and Sisk, W.E. (1984) Incineration of explosives contaminated soil. In: Ref. 50, 203–206.
27. Khan, A.Q. and Thomas, G.A. (1980) In: Ref. 50, 226–232.
28. Various (1985) Conference papers: Biotechnology: new options in land decontamination, Birmingham. European Studies Conferences, Uppingham, UK.
29. Mokveld, F.C.H., Huber, T.F. and Luijben, K.Ch.A.M. (1986) Reclamation of a contaminated soil with a bioreactor. In: Ref. 15, 895.
30. Bewley, R.J.F. (1986) A microbiological strategy for decontamination of polluted land. In: Ref. 15, 759–768.
31. Fredricks, A.N. A case study (of microbial treatment) from Holland. In: Ref. 27.
32. Biotechnica Limited, (1985) Promotional literature. Cardiff, United Kingdom.
33. Landreth, R.E. (1983) Physical properties and leach testing of solidified/stabilised industrial wastes. US Environmental Protection Agency, Cincinnati, (NTIS Order No. PB 83-147983).
34. Anon. (1985) Getting to grips with waste solidification. ENDS Report, (120), 11–13.

35. Ministry of Agriculture, Fisheries and Food (1973) Lime and liming. HMSO, London.
36. Lars-Gunnar, Lindfors, (1980) Reclamation of site of herbicide factory. *Proc. Conf. Reclamation Contaminated Land*, Eastbourne, Society of the Chemical Industry, London.
37. Anon. (1980) Measures for treating soil contamination caused by hexavalent chromium in Tokyo.
38. Sanning, D.E. (1985) In-situ treatment. In: Ref. 14, 91–111.
39. Michelsen, D.A. (1984) The use of microdispersion of air in water for in-situ treatment of hazardous organics. In: Ref. 50, 398–403.
40. Rott, U. (1981) Protection and improvement of groundwater quality by oxidation processors in the aquifer. *Proc. Intern. Symp. Quality of Groundwater*, Noordwijkerhout, The Netherlands. *Studies in Environmental Science*, Vol. 17. (ed. P. Glasbergen,) Elsevier, Amsterdam.
41. Barry, D.L. (1982) Treatment options for contaminated land. Atkins Research and Development, Epsom.
42. Bodocsi, A. Minkarah, I. and Randolph, B.W. (1984) Reactivity of various grouts to hazardous wastes and leachates. *Proc. 10th Ann. Research symp. on Land disposal of hazardous wastes.* US E.P.A. Cincinatti. Report 600/9-84-007, 43–51.
43. Fitzpatrick, V.F., Buelt, J.L., Oma, K.H. and Timmerman, C.L. In-situ vitrification—a potential remedial action technique for hazardous wastes. In: Ref. 50, 191–194.
44. Brown, K.W., Thomas, J.D. and Green, J. (1984) Permeability of compacted soils to solvent mixtures and petroleum products. *Proc. Tenth Annual Research Symposium, Land Disposal of Hazardous Wastes.* US Environmental Protection Agency, Report. EPA 600/9-84-007, 124–137.
45. Anderson, D.C., Gill, A. and Crawley, W. (1984) Barrier-leachate compatibility: permeability of cement/asphalt emulsions and contaminant resistant bentonite/soil mixtures to organic solvents. In: Ref. 50, 131–134.
46. Beine, P.R.A. and Geil, M. (1986) Physical properties of lining system under percolation of waste liquids and their investigation. In: Ref. 15, 863–866.
47. Childs, K. (1985)
A. Management and treatment of groundwater: an introduction.
B. In-ground barriers and hydraulic measures.
C. Treatment of contaminated groundwater.
D. Mathematical modelling of pollutant transport by groundwater at contaminated sites. In: Ref. 14, 141–205.
48. Weinberg, R., Heinze, E. and Forstner, U. (1986) Experiments on specific retardation of some organic contaminants by slurry trench materials. In: Ref. 15, 849–858.
49. Weststrate, F.A. and Loxham, M. (1986) The use of vertical cut off walls in the containment of pollutants. In: Ref. 15, 859–862.
50. *Proc. Conf. Management Uncontrolled Hazardous Waste Sites*, (1984) Washington D.C. Hazardous Materials Control Research Institute, Silver Spring, Maryland.
51. Hass, H.J. (1986) All round encapsulation of hazardous waste in the soil by means of grouting gets and sealing walls resistant to aggressive chemicals. In: Ref. 15, 867–870.
52. Slurry trench construction for pollution migration control (1984) US Environmental Protection Agency, Washington DC, Report EPA—540/2-84-001.
53. Coldewey, W.G. (ed.), (1984) Untersuchungen zur Wasserdurchlaessighkeit bindiger Boden. Investigation of the water permeability of cohesive soils. Mitteilungen der Westfaelischen Berggewerkshchaftskasse, No. 43.
54. Coldewey, W.G. (1984) Measurement of low permeability co-efficients by means of electronic instruments. In: Ref. 50, 584–587.
55. Sullivan, J.M., Lynch, D.R. and Iskandar, I.K. The economics of ground freezing for management of uncontrolled hazardous waste sites. In: Ref. 50, 386–392.
56. Parry, G.D.R. and Bell, R.M. (1985) Covering systems. In: Ref. 14, 113–139.
57. Bradshaw, A.D. and Chadwick, M.J. (1980) *The restoration of land.* Blackwell, London.
58. Coppin, N.J. and Bradshaw, A.D. (1982) *Quarry reclamation.* Mining Journal Books, London.
59. Williamson, N.A., Johnson, M.S. and Bradshaw, A.D. (1982) *Mine wastes reclamation.* Mining Journal Books, London.
60. Dutton, R.A. and Bradshaw, A.D. (1982) Land reclamation in cities. HMSO, London.

61. *Handbook—Remedial Action at Waste Disposal Sites* (1985) US Environmental Protection Agency, Washington DC, Report EPA/625/6-85-006.
62. *Handboek bodemsaneringstechnieken.* (*Handbook of soil reconstruction techniques.*) (1983) Staatsuitgeverij, The Hague, ISBN 90-12-04404-9.
63. Doubleday, G.P. (ed.) (1983) Reclamation of former iron and steelworks sites. Durham County Council/Cumbria County Council.

6 Soil cover reclamations

T. CAIRNEY

6.1 Introduction

Covering unwanted or unpleasant materials with clean soil is a long established historical practice.

When large scale industrialization commenced and produced volumes of unwanted wastes, covering was often seen as the obvious answer to the problem. The wastes were tipped into whatever convenient hollows existed, capped with soil or hardcore, and usually became the foundations for later generations of the industrial activities. Typically, town gas generating plants, which gave rise to large volumes of clinker, ash, spent oxide, tars and other liquors (1), grew on and over their own wastes (Figure 6.1). In other cases a site with over 100 years or more of use might be occupied by several different industries, each of which produced its own discrete waste materials. The effect of this is that many old industrial sites are now underlain by several metres of variably contaminated materials (2).

This approach continued until well after the Second World War, and attracted little or no comment at the time. The benefits of cheaply and easily reclaiming scarce land were seen as self-evident, and any longer term consequences of burying what would now be recognized as chemical contamination were simply ignored.

With the publication of Rachel Carson's book—*Silent Spring*—(3), the more environmentally aware sections of the public came to realize that natural pathways do exist, and that these can concentrate some contaminants to the point where they present direct hazards to people or the food chain which supports our society. Thus the previous easy acceptance of soil cover methods came to be questioned, and well reported failures, such as the Dutch Lekkerkerk soil cover reclamation (Chapter 1), emphasized this view, although the Lekkerkerk failure is now known to have been more a consequence of poor design and bad reclamation control, than of a basic flaw in the soil cover method. Other more successful soil covers (4) were ignored in this criticism.

The difficulty with the soil cover approach is of course that the chemical contaminants still remain on the site, albeit below a thickness of clean cover, and pathways can transmit the contamination to the clean higher surface of the site (Figure 6.2). Thus many workers have found it difficult (Chapter 5) to see soil covers as effective long term remedial solutions in those cases where the contamination is mobile.

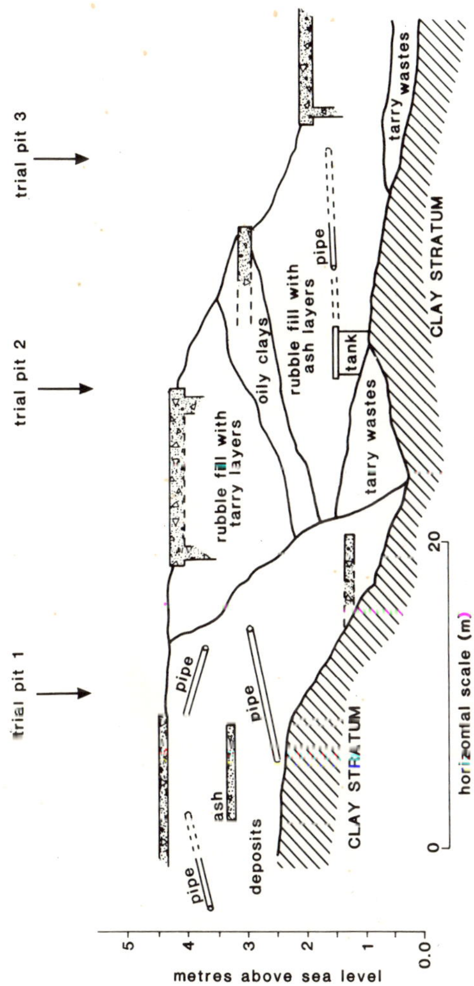

Figure 6.1 Cross-section of part of a gasworks site

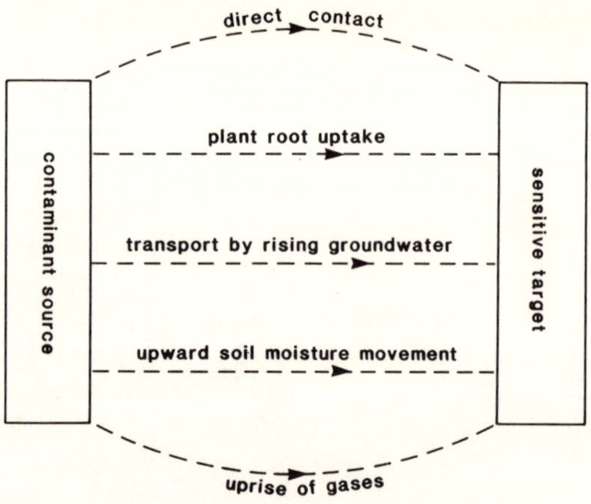

Figure 6.2 Pathways to contaminant migration

Because of these reasonable worries, very few soil cover reclamations (Table 6.1) took place in Britain between 1975 and 1980.

However, in the last few years, an increasing confidence has grown that the pathways can be quantifiably closed on a long term basis to contaminant migration, and (in 1985) to the author's personal knowledge at least 20 British soil cover reclamations are currently being planned or are in progress.

To understand the reasons for this shift in attitudes calls for a clear identification of the performance requirements a soil cover has to meet, and of the methods of properly designing such a reclamation.

Table 6.1 British soil cover reclamations reported pre-1980

Site	*Prior use*	*Proposed use*	*Cover details*	
Locality not reported (5)	Gasworks	Housing	Top soil (225 mm) laid over a variably thick bulk fill, over 'hoggin'	
Thamesmead Arsenal	Gasworks and general tip	Public open area	Top soil	1000 mm
			Clay	300 mm
			Shingle	no data
			Rubble	1000 mm
Beckton Alps	Gasworks	Public open area	Top soil	300 mm
			Clay	1200 mm
			Drainage sand	225 mm
Willow Tree Lane	Gasworks and other waste tip	Housing and public open area	Top soil	150–550 mm
			Porous layer	25 mm
			Hard Core	150 mm
Grandpont Gas works, Oxford	Gasworks	Housing and public open area	Top soil	no data
			PFA	no data
			Clay seal	1000 mm
Swansea valley	Metal Smelting	Amenity open area	Top soil	no data
			Slag	> 1 m thick

6.2 Performance requirements for soil cover reclamations

As indicated in Chapters 1 and 5, the British emphasis is to produce safe environmental conditions at the surface of a reclaimed site, ensuring that its sub-strata are adequate and safe foundations for buildings and structures, and to bring derelict land back into effective use as quickly and cheaply as possible. Generally, the improvement of the quality of groundwater below a site is not a British priority, since most contaminated sites lie in the older industrial areas where groundwater is often so polluted that it cannot economically be used for supply purposes. This viewpoint may, of course, be changed if EEC Regulations are brought in which require groundwater quality improvement to become an intergral part of the overall reclamation (Chapter 5).

On the current British priorities, a soil cover has to

(a) Provide a safe and permanent barrier between the buried contamination and the new higher surface of the site
(b) Give the engineering and environmental conditions needed to ensure that the planned re-use of the site is successful, and
(c) Be as inexpensive as long-term safety requirements will permit.

In more detail, (6), the engineering and environmental performance requirements can be expanded to include:

(i) Minimizing toxicity in the upper layers of the site to levels below those accepted as safe, (7)
(ii) Providing a surface layer which will not become flooded in periods of heavy rainfall
(iii) Controlling any lateral movement of infiltrated rain within the site
(iv) Controlling any tendency for an upward movement of contaminated fluids to occur (from the buried contaminants) in periods of long hot droughts
(v) Preventing or reducing the passage of infiltrating rain down into the buried contaminants to further pollute the local groundwater
(vi) Improving the bearing capacity of the site's soils
(vii) Resisting any subsidence, caused by differential settlement in the weak materials which often underlie such contaminated sites
(viii) Preventing or minimizing the risks of combustion on those sites where the contaminated layers contain appreciable quantities of coal, coke, pitch or other combustible materials
(ix) Controlling the rise of any gases the site may produce
(x) Be resistant to erosion or to slope instability
(xi) Be as free from routine maintainance as possible
(xii) Be able to meet the above requirements in as quantifiable a manner as current knowledge permits.

This last requirement—for a quantifiably safe soil cover—is all too often ignored, but is essential since soil covers will have to be effective for periods of perhaps 25 years (the likely life of a light industrial estate) to 100 years or more (where domestic housing is built on the site), during which more than adequate time is available for a wide range of climatic, environmental and human factors to adversely affect the soil cover's design performance. Present knowledge is inadequate to allow every performance requirement to be safely quantified, although it is possible to make realistic estimates of most of those listed, and the present uncertainty on some performance criteria will inevitably decrease as long-term monitoring (see Chapter 7) produces performance data on particular reclamations.

6.3 The contaminant problem

Chapter 3 describes the more common contaminants encountered on derelict industrial sites, and points out that most are water soluble to some extent. It is these more mobile contaminants, and the gases (see Chapter 11) which also can be present, and which pose the gravest difficulties to the long term effectiveness of a soil cover reclamation.

Equally hazardous substances, such as the pervasive solid tars and pitches which occur on most gasworks and similar sites, are of course a hazard if people, animals, or food crops come into direct contact with them. However, an adequately laid soil cover, at least, will be a physical barrier preventing such

Table. 6.2 Chemical qualities of solid and water samples, in adjacent trial pits, gasworks site, N.E. England

	Trial pit 1		*Trial pit 2*		*Trial pit 3*	
	Solid sample	*Water sample*	*Solid sample*	*Water sample*	*Solid sample*	*Water sample*
pH	7.3	8.0	4.1	2.7	7.7	6.3
Chloride	ND	44	ND	210	ND	28
C.O.D.	ND	50	ND	72	ND	35
Total cyanide	4	1.0	<2.5	3.3	<2.5	0.15
Free/simple cyanide	2.7	<0.05	<2.5	<0.05	<2.5	<0.05
Toluene extract	3200	72	3200	350	900	60
Total phenols	<2.5	<0.05	<2.5	<2.5	<2.5	<0.05
Sulphate (SO_4)	800	240	10 000	2900	300	380
Sulphide (H_2S)	2	<0.02	4	<0.02	2	0.4
Total arsenic	24	<0.005	18	<0.005	12	<0.005
Total lead	430	<0.03	280	0.08	60	<0.03
Total magnesium	ND	45	ND	100	ND	34
Total zinc	140	<0.03	320	12	31	<0.003

ND = Not determined
< = Less than
All values expressed in milligrams of contaminant per kilogram of soil, or milligrams of contaminant per litre of water sample, except pH which is an absolute value.

Table 6.3 Range of solid and water qualities, tar works site, S.W. England

	Solid sample range	*Water sample range*
pH	11 – 7.5	8.1 – 7.2
Sulphate as SO_3	72 000 – 200	6400 – 200
Phenols	87 – 2	22 – 0.25
Toluene extract (%)	9.7 – 0.19	0.9 – 0.001
Copper as Cu	12 400 – 24	0.96 – 0.09
Nickel as Ni	129 – 32	1.25 – 0.09
Zinc as Zn	3150 – 62	0.828 – 0.104
Cadmium as Cd	8.2 – 1.6	0.250 – 0.006
Lead as Pb	1550 – 10	0.184 – 0.053

All values expressed as on Table 6.2.

contact, and thus these less mobile contaminants are seen as minor hazards in the context of soil covers.

The range of water solubility of the commoner contaminants can be proved in laboratory tests (Chapter 4), but are perhaps best indicated by analyses of soil and water samples from actual site investigations (Tables 6.2 and 6.3). Such information reveals that high levels of mobile contaminant can exist, and, if uncontrolled, might feasibly reduce the long term safety of a soil cover reclamation.

Before it can be concluded that mobile contaminants do present a long term hazard, it is essential to define the various pathways which might allow the contamination to affect the surface of the reclaimed site, and to decide whether these pathways can be effectively closed.

6.4 Pathways for contaminant migration

Five pathways (Figure 6.2) are normally identified:

(a) If workmen or residents excavate through the soil cover and come into direct contact with contaminated materials. As Chapters 4 and 5 point out, this could occur as residents decide to extend their homes or to build additional structures
(b) If deep rooted plants can penetrate through the soil cover into the buried contaminants and then concentrate hazardous substances in leaves or fruit
(c) If gases are produced deeper within the site, and these can percolate upwards to affect the vegetation on the site surface, or collect beneath buildings to pose toxic or explosion risks
(d) If the contaminants are water soluble and can be lifted up by uncontrolled groundwater, and
(e) If the soil cover can allow significant quantities of polluted soil water to rise upwards to heights well above the groundwater level, in periods when the upper layers of the site are dried out by extreme droughts.

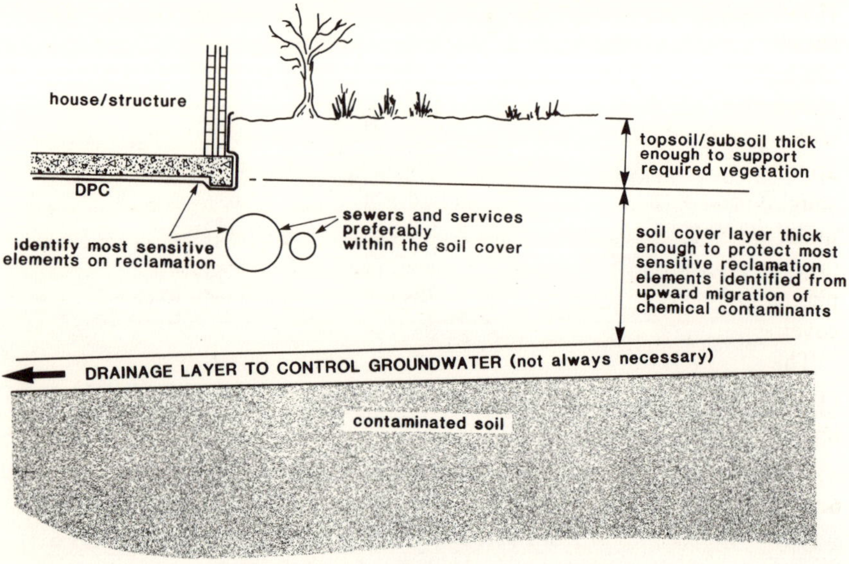

Figure 6.3 Basics of soil cover reclamation

The problem of workmen and residents excavating through a soil cover is solvable largely by sensible administration.

Buried services (gas, electricity and water supplies) should always be laid within the clean soil cover (Figure 6.3), even if this means laying a rather thicker cover than the design method (section 6.7.) requires. Sewerage, however, usually presents a greater difficulty, since the gradients necessary for effective self-cleansing flow in sewers may imply that the sewers have to be laid within the contaminated soils, below the cover blanket. In such cases, the most reasonable solution is to have all sewerage trenches over excavated, and the pipework surrounded by a zone of clean permeable material to ensure that workmen engaged in any later repairs to the pipework do not directly encounter contaminated materials, or have to work in polluted groundwater. Such sewerage trenches should be in well defined reservations, and accurate records of all buried services should be kept by all the public utility bodies. These records should also include details of the contaminant hazards likely to occur beneath the soil cover, together with site safety instructions similar to those detailed in Chapter 8.

The problem of local inhabitants excavating through the soil cover is less easily resolved over the likely life of (say) a domestic housing estate, and the sole practical safeguard is to ensure that the soil cover is always thick enough to allow for such activities.

The particular difficulty of establishing plants over a soil cover reclamation is part of the material covered in Chapter 10, and it would be inappropriate here to discuss how various plant species respond to chemical contaminants.

However, the general rule for soil covers is simple. No deep rooted species should be established over a soil cover reclamation. Where gardens or allotments are required, an adequate thickness of top and sub-soil should always be placed *above* the soil cover layer to allow the plants to survive, without their roots having to penetrate deep into the site. The advice of appropriate specialists—botanists or landscape architects—should always be sought before any large scale revegetation of a soil cover reclamation is ever undertaken. For those situations where amenity tree planting is necessary in (say) a domestic housing estate, a sound policy is to establish the trees in planters with solid concrete bases to prevent root penetration into the soil cover layer.

The generation of gases and the methods of reducing the hazards of these are discussed in Chapter 11, and will be touched on in the examples (section 6.7.) given later in this Chapter.

6.5 Groundwater and soil water movements

6.5.1 *Introduction*

Of the five ways in which buried contamination can affect the higher levels of a soil cover reclamation, water-borne contaminants are the most serious, and it is against such cases that soil cover design is most severely tested.

That many industrial by-products are water soluble is not surprising. Most such substances were formed by conditions (reducing atmospheres or high temperatures) very different from those of the natural environment. Thus they tend to be sensitive if exposed to air and water at the lower ambient temperatures of today's environment, and groundwater which comes into contact with them thus often becomes highly contaminated (Tables 6.2 and 6.3).

The upward migration of such contaminated water and the methods of minimizing this are thus the major concerns when designing a soil cover blanket.

Two separate transportation mechanisms (that occur at quite different seasons of each year) exist—the cyclic rise of groundwater to flood the soil cover layer, and the capillary rise of moisture to heights well above that reached by the groundwater level—and each has to be combatted by the design of the soil cover reclamation.

6.5.2 *Groundwater cyclic movements*

Groundwater is, of course, no more than that proportion of rain or snowfall that is able to infiltrate or percolate into the ground. Flatter topographic gradients and a ground surface that is permeable to the entry of water encourage this infiltration.

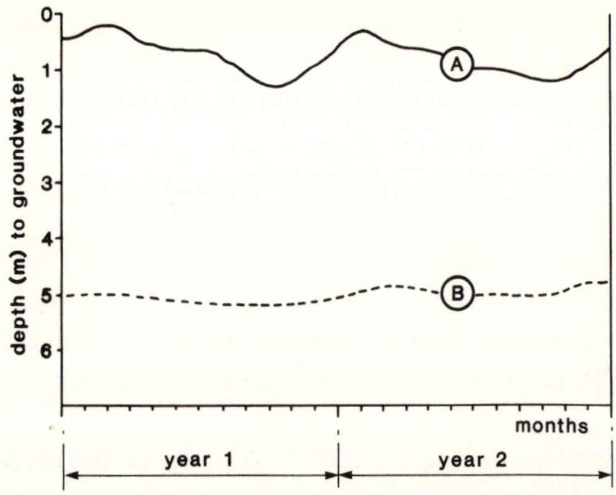

Figure 6.4 Typical annual water level change in (*A*) a perched groundwater and (*B*) a deeper groundwater

Infiltration rates are also affected by evaporation from the site surface, and by the amount of soil moisture abstracted by plants for their biological uses. In winter, when evaporation is low and when most plants are dormant, a much higher infiltration occurs, and, because of this, groundwater levels generally show a cyclic and seasonal pattern (Figure 6.4), with the shallowest depth to groundwater occurring at the end of the winter months (February to April), and the deepest groundwater being in the Autumn (September to October).

Apart from these seasonal factors, groundwater levels can also vary in sympathy with those of nearby open bodies of water, particularly if these open stretches of water are subject to tidal effects.

The actual amount of movement of groundwater levels over a year depends on a variety of local factors, but it can exceed one metre, and cases where much greater variations occur are not uncommon.

The effect of this on a soil cover reclamation can be serious. If groundwater can rise into the soil cover layer, then the flooded portion of the soil cover is no longer effective. In more extreme cases, the groundwater might reach the new surface on the site to form springs or pools of contaminated water. In either case, the soil cover barrier has failed in its design role.

Thus it is important that an early sub-surface investigation of the site should prove the position of the groundwater table, the quality of the groundwater, and that enough monitoring should be carried out to establish any cyclic rise and fall of the groundwater level.

Despite the self-evident nature of the above remarks, it is surprising how few site investigation reports adequately define the depth to groundwater level at the time of the drilling investigation, and how rare it is to be given any detail on the possible range of cyclic variations in the depths to groundwater. Presumably this occurs only because the site owners, who commission the investigations, omit to require this information from their contractors.

If the groundwater's position is properly established, and if its maximum height in the ground can be estimated, then a variety of standard drainage techniques can be used to ensure that the groundwater can never rise above some chosen level and flood the soil cover layer. Such drainage works should be installed prior to the placing of the soil cover layer.

6.5.3 *The capillary rise of polluted soil moisture*

If the groundwater is restrained to some chosen depth below the soil cover, it might be thought that no possibility of water rise up through the soil cover layer can exist.

This view however overlooks the ability of dry soil materials to lift thin capillary threads of moisture from deeper and wetter horizons. The range of such possible upward movement (for some soil types) can be in excess of 3 m, a value greater than the thickness of most soil cover layers, and so the capillary movement of soil water is a particularly important aspect of soil cover reclamation design.

Whilst the concept of moisture moving upwards in a soil column appears strange, analogous and more familiar phenomena are the occurrence of 'rising damp' in buildings, and the action of a wick in an oil lamp.

The two factors that control any movement of moisture in a porous medium (such as a soil) are the force that creates the potential for the flow, and the conducting ability of the porous medium, that governs how much flow actually occurs.

In a partially dried out soil, the controlling force is the difference in soil suction values (ψ) between different levels in the soil profile. Soil suction is no more than the ability of a dry soil to take moisture from wetter adjacent layers, and this arises because the water molecules, that form directly on the surfaces of a grain of soil, are held by very strong adhesion forces, and can attract further water molecules to join them. The drier a soil, the greater is its ability to attract moisture from adjacent soils, and the higher is its soil suction value (8). Conversely a fully saturated soil (at the groundwater table) has no suction ability, since all its pores are already full of water.

In the rarer very hot dry summers that affect the UK (such as 1976), the upper layers of a soil profile become almost totally dried out, whilst deeper in the soil profile, the soil is progressively wetter until at the groundwater level the soil pores are all full of water. Thus a marked difference in soil suction exists across the profile (Figure 6.5), and the greater this difference, the larger is the potential for an upward migration of moisture (and whatever dissolved contaminants this moisture may have) from the groundwater level.

The actual rate of upward migration of soil moisture depends on the ability of the particular soil to pass a flow of water. This ability is technically defined as the soil's hydraulic conductivity (or permeability)—'K'—and is usually expressed in cm^3 of water per day passing through each cm^2 of soil measured

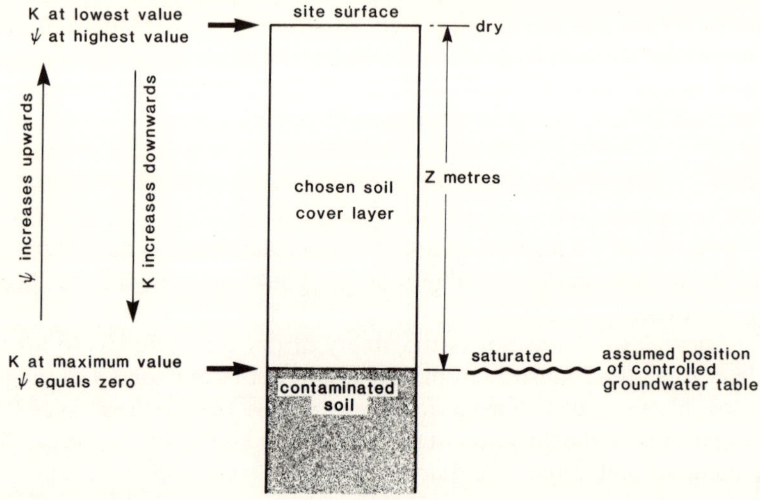

Figure 6.5 Factors which affect upward migration of soil water

at right angles to the direction of moisture movement. When soils are fully water saturated, the more granular types (sands and gravels) have much higher conductivities than do fine grained clays and silts, and it is for this reason that drainage layers and soak-aways are invariably made of coarser grained materials.

However, these coarser soils have large soil pores, which drain easily when the soil dries out, and which then become airlocked. Thus when a sand or a gravel becomes drier, its hydraulic conductivity value decreases rapidly (Table 6.4). Conversely the finer grained clays and silts (which have quite low conductivity values when fully water saturated) have very fine soil pores in which capillary forces prevent easy drainage. Thus the decline of hydraulic conductivity in a partly dried clay or silt is much less than that which occurs in a sand or gravel at an equivalent stage of dryness, and—at particular states of desiccation—the finer grained materials actually can have higher conductivity values than do the sands and gravels. (Table 6.4).

Table 6.4 Comparative physical properties with changes in soil water content

Saturation Value of Soil%	*Coarse gravelly sand*		*Fine beach sand*		*Estuarine clay*	
	Permeability cm/day	*Suction cm H_2O*	*Permeability cm/day*	*Suction cmH_2O*	*Permeability cm/day*	*Suction cm H_2O*
100	2100	0	310	0	0.65	0
80	174	6.5	117	31	0.46	210
60	6.5	10	7.4	50	0.02	430
40	0.0001	65	0.04	79	0.01	1137
20	10^{-8}	84	10^{-6}	310	4.3×10^{-3}	12375

From the above comments, it is obvious that little can be done about the difference in soil suction values across a soil profile, whose surface is open to the atmosphere, since these are controlled by climatic conditions. However, the amount of upward flow that actually takes place can be minimized by choosing a soil cover material which is incapable of conducting any large rates of flow. Thus the design choice is to select that cover material which is least liable to permit upward migration of polluted soil moisture.

The rate of upward soil moisture movement is described by the equation

$$V = K(\psi)\left[\frac{d\psi}{dZ} - 1\right] \qquad (1)$$

where V = the flux (i.e. rate of flow) in cm^3/day through each cm^3 of soil surface

$K(\psi)$ = the value of the soil's hydraulic conductivity that occurs at particular state of soil dryness

ψ = the suction value developed at whatever the soil's state of dryness

and Z = the thickness of soil cover above the groundwater table.

To use this equation, one has to know how a soil material's hydraulic conductivity and suction values vary as the soil is dried out (Table 6.4). Various specialized laboratories are able to conduct such tests and rapid and accurate test methods (9) are now available.

Even when the variation in conductivity and suction values with water content are known, Equation (1) is difficult to use manually since at each level in the soil profile the soil moisture content is different, and so the soil's conductivity and suction values change at every height above the water table. Computer assistance is generally necessary, and models such as that devised by Bloemen (10) allow the upward flow rate that passes each level in the soil profile to be calculated. Bloemen's model is simple to use and allows the flow rate to be calculated for whatever the suction difference that exists across the soil profile. As suction difference is no more than a measure of the state of soil surface dessication that will occur under particular drought conditions, the model thus allows a designer to consider the effects of various historical drought conditions on the proposed soil cover.

Emphasis has been given above to the suction and hydraulic conductivity concepts, since these will be unfamiliar to most readers. However, Equation (1) also makes it clear that the depth (Z) down to the controlled groundwater level is also crucial, and the greater this depth, the smaller will be any upward movement of polluted soil water. This gives added emphasis to the importance of controlling the maximum level to which groundwater can rise by installing an adequate site drainage, before the soil cover is laid.

The Bloemen model is capable not only of providing accurate upward flow rates through a simple one-material soil cover, but can also quantify 'sandwich' soil covers which include up to 8 different soil layers. Such 'sandwich' covers are sometimes necessary (11 and 12).

Another implied assumption in the above discussion is that the uppermost layers of a soil cover will largely dry out in a long hot drought and so promote a meaningful upward migration of polluted soil moisture from the buried contaminants. Whilst this is indeed so for those parts of a soil cover that are open to the atmosphere, other parts of the reclaimed site will be blanketed with buildings and roads. In these cases (13), the soil cover will be less dessicated and a much smaller tendency for upward moisture flows will develop. This, of course, gives a useful extra factor of safety to those more sensitive reclamations, where much of the site surface is built over.

Capillary upward movement of soil moisture is a slow process and one that will only be significant in those rarer and hotter drought summers. Thus, whilst capillary movements are important (especially if a reclamation has to provide a safe permanent barrier for many tens of years), it must never be forgotten that a soil cover layer (in Britain, though less so in more arid areas) will be exposed to the normal downward drainage of moisture for the greater part of each year, and that groundwater control and drainage is therefore of greater practical significance.

6.6 Choice of soil cover materials

Whilst it is possible (6) to consider cover materials which include

(a) The entire range of natural soils
(b) Natural soils whose properties are amended and improved by the addition of (for example) cement or pulverised fuel ash ('fly ash')
(c) Synthetic materials, and
(d) Wastes such as slags, building rubbles or pulverized fuel ashes.

British experience is so far limited to the use of natural soils and to some of the more appropriate waste materials.

The modification of natural soils' properties is an attractive idea, but the additional costs of this, and the practical difficulties of attaining a uniform degree of modification over a large site's area have so far prevented any application of the method.

The use of synthetic materials in experimental settings has been confined to a few cases, and these have suggested that many man-made materials are inadequately robust to give the long term security that a soil cover must possess (14 and 15). Thus their use on a practical reclamation cannot as yet be recommended.

Whilst the range of natural soils and available waste materials still appears to offer a wide choice to the designer of a soil cover reclamation, practical factors invariably reduce this theoretical choice to a handful of available materials. The reason for this can be understood if it planned (say) to reclaim a 4.530 hectare site by covering it with a 1.30 m thick clean soil layer. The

Table 6.5 Sieve analysis results

Material	*Clay and silt* (%)	*Sand* (%)	*Gravel* (%)	*Organic content*
Silty gravelly sand	20	45	35	0
Sandy silt	36	62	0	0

volume of material needed (58 890 m^3) is such that only very few local sources will be available, and if materials have to be transported for more than a few kilometres to the site, then the costs will quickly become uneconomic. Although the costs of bulk fill materials vary with local factors, it would not be unusual in the North of England to expect to pay £2.50/m^3 for a locally available soil, thus making the material costs for the example considered £147 225, to which must be added the costs of laying and compacting the soil cover.

Thus the normal situation is for a choice of perhaps only 3 or 4 local materials to be available, and the problem is to select the best and decide how thick a cover is needed to give the required safety.

From some sources it might be believed that a clay cover is desirable, whilst other publications (16) argue strongly for the use of granular materials (coarse sands or gravels) and advise against the inclusion of clays or fine silts. To resolve this debate, the designer should consider the performance requirements the soil cover must possess (section 6.2), refer to data such as that given in Table 6.4, and to standard texts (17) on the propensity of clay soils to shrink, crack, swell and heave in different climatic conditions.

Clayey materials have properly been recommended (18) as the cap materials for old land fill sites where the prime design feature is to minimize rainfall infiltration, which would add to the leachate problems of such tips. However,

Table 6.6 Permeability—water content—suction variations

Silty gravelly sand			*Sandy silt*		
Water Content (%)	*Hydraulic conductivity cm/day*	*Suction on pF scale*	*Water content* (%)	*Hydraulic conductivity cm/day*	*Suction on pF scale*
13 saturated	18.4	0	25 saturated	5.04	0
8.5	1.2	1.0	20	1.13	1.5
6	3.4×10^{-2}	1.8	15	3.4×10^{-1}	1.7
4	1.7×10^{-3}	2.75	10	3.0×10^{-3}	1.9
2	7.1×10^{-5}	3.5	5	1.0×10^{-3}	2.5
1	1.0×10^{-6}	4.0	2.5	2.0×10^{-4}	4.0

(pF is the logarithmic suction scale which records a suction of 1000 cm of water as a pF of 3)

Table 6.7 Predicted upward flow of soil moisture

Thickness of cover blanket (cm)	*Flow ($cm^3/day/cm^2$ of surface) reaching top of cover blanket made up of* (*a*)	(*b*)
	Silty gravelly sand	Sandy silt
20	0.0400	0.4500
40	0.0115	0.0800
60	0.0058	0.0300
80	0.0035	0.0120
100	0.0022	0.0070
120	0.0015	0.0040
140	0.0012	0.0030
160	0.0010	0.0019
180	0.0010	0.0015
200	0.0010	0.0010

where developments are to take place above a soil cover, clays are generally poor foundation materials, can cause surface flooding problems, and promote very much greater upward movements of polluted soil water (see Tables 6.7 and 6.8) in long hot droughts. Additionally the thickness of a clay or silt rich material to reduce the upward migration of soil water to some acceptable level is much greater than that required for a coarse granular soil (Table 6.9), and so clay covers will be more expensive, as well as giving poorer foundation and drainage conditions.

Thus it is generally better to use the coarsest granular soil cover material that is available, and to limit the use of clayier materials (where these have to be used) to layers deeper in the soil cover where they can be protected from desiccation and freezing.

Table 6.8 Highest contaminant levels recorded in ground water samples

Chemical oxygen demand	16 000
Chlorides	2400
Cyanide (total)	32
Cyanides (free and simple)	0.32
Toluene extract	5400
Total phenols	23
Sulphate as SO_4	7900
Sulphide as H_2S	280
Total arsenic as As	0.23
Total lead as Pb	0.60
Total magnesium as Mg	340
Total zinc as Zn	13

All values expressed in milligrams per litre.

Table 6.9 Concentration of sulphate added by flow to top surface of soil cover blanket

Thickness of cover blanket (cm)	*Silty gravelly sand cover layer*	*Sandy silt cover layer*	*50 cm of silty gravelly sand covered by a layer of sandy silt*
40	6148 mg/kg	42 753 mg/kg	—
60	3099 mg/kg	16 032 mg/kg	3206 mg/kg
80	1873 mg/kg	6413 mg/kg	2086 mg/kg
100	1176 mg/kg	3741 mg/kg	1873 mg/kg
120	804 mg/kg	2091 mg/kg	1552 mg/kg
140	641 mg/kg	1604 mg/kg	1068 mg/kg
160			855 mg/kg
180			641 mg/kg
200	535 mg/kg	535 mg/kg	535 mg/kg

The third column in Table 6.9 was included in case inadequate quantities of the silty, gravelly sand were available to cover the whole site to the required thickness.

6.7 The design of soil cover reclamations

6.7.1 *Introduction*

The various themes developed above are perhaps best illustrated by the details of particular reclamations.

6.7.2 *Gasworks site, N.E. England*

This had been used for town gas generation for over a century, and the waste volumes of demolition debris, ashes, clinker, spent oxides, and foul lime finally filled and levelled the original clay-floored valley, on which the first generation of gasworks had stood.

When gas production ceased, the site was reasonably level, extended over 8 hectares, and was underlain by rather more than 2 m thickness of contaminated wastes, in which a near surface perched groundwater table exists. The solid and water samples obtained in the site exploration showed the usual variability that typifies such sites, but high contamination levels were widespread (Table 6.2).

The decision to use a soil cover reclamation was forced by the realization of the costs and difficulties of excavating and removing more than 189000 m^3 of contaminated materials, and by the identification of an artesian aquifer just below the original clay floored valley. Water pressures in this aquifer are such that removing the soil loads above it (by excavating the gasworks wastes) could have allowed the artesian water to break through and flood the lower levels of the site.

Once a soil cover was chosen, the problem was to identify the most sensitive elements proposed for the land's re-use. Although planning had not been fully completed, it was decided that the site should not house domestic gardens, allotments or any deeper rooted revegetation, and that it should be reclaimed to permit a range of light industrial uses, domestic dwellings (without gardens) or amenity activities.

Given this range of end-uses, it was easy to identify the building foundations (and their associated buried services) as the most sensitive elements, since the houses might well have to be occupied for up to a century, a life longer than that likely for light industrial premises.

A design based on protecting house foundations for a century of use was thus chosen, and local meteorological records revealed the wettest and driest periods likely to occur in such a time.

Monitoring of the perched groundwater had been carried out for six months, and from this (and the historical meteorological records), it became apparent that the groundwater could reach the existing site surface in wetter periods, and might well be able to rise up into whatever soil cover was laid over the site.

Thus the initial reclamation activity was to construct rubble filled drains over the site to ensure that groundwater was never able to rise above the site's present surface levels. The opportunity given by this drainage programme also allowed all concreted areas of the site to be broken up, and various underground vats of tars and other liquors were located, and emptied by specialist contractors.

When the site was being drained, efforts were made to locate suitable locally available soil cover materials, and two soil types were finally found and analyzed (Tables 6.5 and 6.6).

The earlier meteorological analyses had indicated that in a 100-year period, the site could expect to suffer a 100-day long drought, which would so dry out exposed sandy soils that suctions, equivalent to 1000 cm of water, would occur.

Thus the stage had been attained at which the design drought was known and the properties of the available soil cover materials were available.

The Bloemen model was then used to quantify what volumes of soil water could rise up through soil covers of two materials, and of various thickness from 20 cm to 200 cm. (Table 6.7).

This last table revealed, as could have been assumed from Table 6.5, that the material with the higher proportion of clay and silt gave the higher upward migration rates of soil water.

The information in Table 6.7, as such, is not in a form that allows a determination of the effectiveness of the different soil covers, and it is always necessary to decide the design chemical quality of the rising soil water.

This was done by identifying, for each analyzed chemical parameter, its highest value in the 58 groundwater samples the site investigation had yielded, and then testing the soil covers against a water which contains all these worst qualities (Table 6.8). Quite obviously this introduces a marked element of over-design, since no individual groundwater attained anything near the contamination levels listed in Table 6.8, but this process appears necessary since there could be particularly poor groundwater qualities in areas of the site left undrilled in the site exploration programme.

The method of converting the upward flow rates, detailed in Table 6.7, into the concentrations of particular contaminants added to various levels of the soil covers, (e.g. the sulphate concentrations given in Table 6.9) is outlined in Appendix I.

Only the sulphate concentrations are listed in Table 6.9, since sulphate posed the greatest direct hazard to the house foundations, but other evaluations were carried out to check that the concentrations of all the other soluble contaminants (see Table 6.8), which would also be lifted into the soil cover by the same rising soil water in the 100-year design drought, would be less than any of the hazardous levels.

Having obtained the data, the final decision was to choose the thinnest possible soil cover that would give the required site safety, and after a consideration of the ICRCL Guidelines (7) and the standards developed by the Greater London Council (see Chapters 1 and 4), it was recommended that if a 100 cm thickness of the silty gravelly sand always lay beneath house foundations, then the reclamation would be safe.

It is worth noting that the figures listed in analyses such as Table 6.9 contain real, though currently unquantifiable, safety factors. No note was taken, in deriving Table 6.9, of any improvement of the soil water (e.g. by cation exchange) as it rose through the soil cover, and, as mentioned above, the design soil water quality is far worse than that of any groundwater known to exist on the site.

The final stage of the design was to ensure that sub-soil and top soil could be laid over those parts of the site where a grass cover might be required without adversely affecting the long term efficiency of the soil cover layer. This required no more than applying the established (19) methods of checking the particle grading curves of the cover material against those of the vegetable soils, and designing a filter layer of medium sand whose particle sizes would prevent any top soil material being washed down by infiltrating rainfall to clog the pore spaces in the soil cover blanket. If this had not been done then, as the soil cover silted up over the years, its suction characteristics would have increased to the point where it was able to promote much greater upward movements of polluted soil water than those detailed in Table 6.9.

What has perhaps not been stressed above is that the testing of the possible soil cover materials (to establish their suction—water content—hydraulic conductivity properties) took place at the bulk density values likely to give the greatest site bearing capacity values, and when the soil cover was laid site control was necessary to ensure that these densities were actually achieved.

If this particular soil cover is assessed against the performance criteria listed in section 6.2 it will be seen that it:

(i) Minimizes the toxicity levels to acceptable values
(ii) Provides good surface drainage conditions
(iii) Controls (via the rubble drain system) lateral water movements in the site
(iv) Minimizes upward movements of soil water in extreme droughts
(v) Reduces the passage of infiltrating rainfall down to the buried contaminants, since the rubble drains discharge their water to a local river and do not permit it to pass into the contaminated waste layers
(vi) Improves the site bearing capacity
(vii) Resists subsidence, in that the cover is well compacted and is elastic enough to take up minor settlements in the underlying gasworks materials
(viii) Is resistant to erosion
(ix) Is designed to overcome silting up and so should be free from the more obvious maintainence problems
(x) Has been quantified and contains a factor of safety, though current knowledge is inadequate to allow a precise definition of this factor.

Neither gases nor combustion risks were significant hazards on the site.

6.7.3 *North London gasworks site*

The above example is somewhat atypical in that complete design freedom existed. More commonly, a designer is constrained by the levels of surrounding roads and land, or sometimes by the requirement to use one particular cover material. In such cases, the main interest is to check the level of safety that the soil cover will give, and to decide whether or not the planned re-use is acceptable.

The North London site is similar in area and in past industrial uses to that detailed above, and presumably once had similar contamination levels. Currently, however, the site is patchily contaminated, with high contamination (especially sulphates, phenols, combustible materials) on its southern half, and much cleaner conditions at its northern boundary with the adjacent river. This pattern appears due to the local groundwater which lies at 3 m below surface on the site's southern margins and then dips steeply northwards to drain to the river. The greater thickness of dry ground on the northern half of the site appears to have allowed water soluble contaminants to be washed out of the site's gasworks wastes.

Combustible materials—high concentrations of coal and coke deposits—occur and fire risks of these were seem as the major difficulty to a safe reclamation.

The planned re-use of the land was for light industrial factories, for which a 25-year design life was specified.

Because of adjoining road levels, the soil cover could be no thicker than 1.2 m. Additionally, the soil cover had to be composed of pulverized fuel ash (pfa) since:

(a) This was available on an adjoining power station stock yard, and
(b) The soil cover's prime function was seen as impeding air entry into the ground to reduce any risks of the coal and coke layers burning, if they ever were heated.

Normally pfa, a silt-like material able to generate very high suctions when dried out, would not be recommended for a soil cover, but in this case the very detailed groundwater monitoring, which included records of the 1976 extreme drought, showed that the cover actually above the local groundwater in a drought would be 1.2 m of pfa underlain by at least 3.0 m of dry silty gas works fill.

The design process outlined in section 6.7.2 was thus carried out, using a design suction of 500 cm of water, (the level to which clay materials would be dried out in a 25-year drought), and revealed that no soluble contaminants could be carried up to the level of the factory unit foundations.

There was, of course, no need in this case to install groundwater drainage controls, since the groundwater (even in a wet winter) was known never to have risen to within 2 m of the original site surface. Drainage of the surface of the reclaimed site was, however, necessary since the compacted pfa cover had too low a permeability to dispel flooding.

The interesting feature of this particular reclamation is not the use of the pfa cover (which by itself would not have prevented upward movements of soluble contaminants) but the advantage given by the 3.0 m of dry silty ground in preserving the safety of the reclamation. The general point this makes is that soil covers are most severely tested when a polluted groundwater can exist within 3 m of the final surface level of the reclaimed site.

6.7.4 *Tar works site, S.W. England*

This site, though smaller than that outlined in section 6.7.2, presented a similar contamination range, (see Table 6.3).

The actual reclamation (which included a soil cover layer composed of a very coarse and permeable (20 mm./second) gravel, which also acted in part as the site's groundwater control) offers little additional insight to the examples already quoted, and the site is included here because of its methane gas problems.

After the initial trial pit site exploration, which proved the thickness and distribution of solid wastes and the quality range of the polluted groundwater, a series of six boreholes was sunk specifically to determine whether gas emissions would pose any hazard. Of these holes, only one initially produced any high concentration of methane, and this declined over a 3-month period,

Table 6.10 Gas monitoring, S.W. England tar works site

Monitoring period (weeks after)	*Field measurement of flammable gas (as % LEL)*	*Laboratory determination of methane as (a) % gas*	*(b) % LEL*
1	60.0	2.350	47.0
3	30.0	1.200	24.0
8	2.25	0.019	0.38
13	1.25	0.015	0.30
18	10.0	0.890	17.80
20	2.50	ND	ND

ND = Not determined

only, however, to rise abruptly in the fourth month of monitoring (Table 6.10).

As this pattern of changing concentrations took place without any meaningful changes in the local groundwater level (see Chapter 11), and in periods when no freezing of the surface occurred, it was obvious that an unexplained (and so potentially hazardous) situation existed.

To cope with this, without delaying the site development, it was decided to use the very permeable soil cover layer as a gas-collecting medium, which would be passively vented to the final land surface (at selected safe locations), by installing vertical chimneys of the coarse permeable gravel up to the final surface. A cover of less permeable fill material, that had already been chosen for laying above the coarse soil cover (to make up site levels to the required contours) was tested and was found to be an adequate cap layer to retain any gases in the soil cover layer and guide them to the vent chimneys.

In addition, all heavier structures are to be founded on piles carried down to the underlying bed rock, and particular care is being taken to gas-proof the interfaces of the piles and the surrounding contaminated soil to minimize any possibility that the piles will act as uncontrolled gas channel ways.

Monitoring of methane levels was accepted as an on-going obligation for this site.

6.7.5 *Waste disposal tip, N.E. England*

This linear 5.6 hectare site flanks the banks of the River Tyne and is adjacent to a heavily populated area where amenity open spaces are scarce. Thus the intention was to cap the hummocky topography, apparently composed of demolition rubble, with a suitable depth of top soil and sub-soil and establish a grassed amenity park.

Site investigation however proved necessary when surface deposits of asbestos were found, and this process proved that amongst the demolition wastes are coal combustion residues, which locally give rise to quite high levels of toxic metals, arsenic, sulphates, phenols and tars.

The exploration found that the wastes could be up to 3 m thick, and that beneath them was a thick deposit of coarse gravel. The local groundwater, whilst contaminated with pollutants from the waste layer was found to lie deep in the gravel, and everywhere was more than 5 m below the site's present land surface.

The most dangerous contaminants on the site were, however, the pervasive asbestos deposits and the local zones of near-solid tars. Both, of course, are insoluble materials.

In this case, a fully designed soil cover to cope with either groundwater transportation or with upward drought induced soil water movements was quite unnecessary, and the problem was to encapsulate the dangerous asbestos and tar wastes.

This was achieved by covering with a well compacted layer of the local glacial clays, above which sub-soil and top soil were placed in thicknesses which were adequate to allow grass and shallow rooted shrubs to survive.

6.8 Summary

It is only worthwhile undertaking the task of a full design of a soil cover in those cases where a near surface groundwater occurs and that water is polluted. In other cases, where insoluble contaminants are the problem, a simpler capping with a suitable soil will probably be adequate, provided that the cap's prime functions are clearly defined and met.

If a full soil cover design is required, then it is imperative that an adequate site exploration takes place to provide as full information as possible on the groundwater, and on its chemical and its level variations with time. The design will also require a complete laboratory testing of the possible cover materials, and an analysis of local climatic factors. Inevitably this process will take several weeks and call for specialist experience.

The design method outlined above allows the soil cover to be quantified to protect whatever is seen as the most sensitive element planned for the reclamation, and despite the greater complexity of the upward moisture movement part of the design, it should never be forgotten that drainage to control the more common case of downward percolation of rainfall is crucial.

Soil covers can also include the range of methods (detailed in Chapter 11) to reduce gas emission problems.

Soil cover techniques are undoubtedly cheap and rapid to install, and require only established engineering methods and equipment. However, they inevitably leave the contamination, that made the site initially unsafe, still in place, and so particular care is necessary to close any feasible pathways by which this contamination could migrate.

Long term monitoring (Chapter 7) should thus be accepted as an inevitable duty arising from soil cover reclamations.

Appendix: Calculation of the contaminant concentration added by the capillary rise of soil water*

Figure 6.6 shows a horizontal strip in a soil cover. This has its lower level at h metres above the groundwater, and its upper surface at the slightly greater height of $h + \mathrm{d}h$ metres.

Bloemen's computer model (see Table 6.7) is used to determine the rates of upward movement of soil water (the fluxes) that would rise through different thicknesses of the chosen soil cover. Such values, when expressed graphically (Figure 6.7), make it apparent that the flux just reaching the base of any horizontal strip in the cover layer will always be greater than that which just reaches its upper surface.

Calculation of the weight of contamination (mg) contained in any flux is by the expression

$$\text{Weight} = c \times \rho \times V \times d \tag{6.2}$$

Where,

$c =$ the chosen concentration of the contaminant in the local groundwater. (mg/kg)
$\rho =$ the density of the groundwater (kg/m^3)
$V =$ the flux, but expressed in m^3/day through each m^2 of soil cover surface.

and

$d =$ the number of days for which this flux persisted.

Since a greater flux will enter the base of the horizontal strip of soil cover than that which will leave through its upper surface, the weight (mg) of contaminant that will be added to the soil cover will be:

$$c \times \rho \times d \times (V\ \text{base} - V\ \text{top}) \tag{6.3}$$

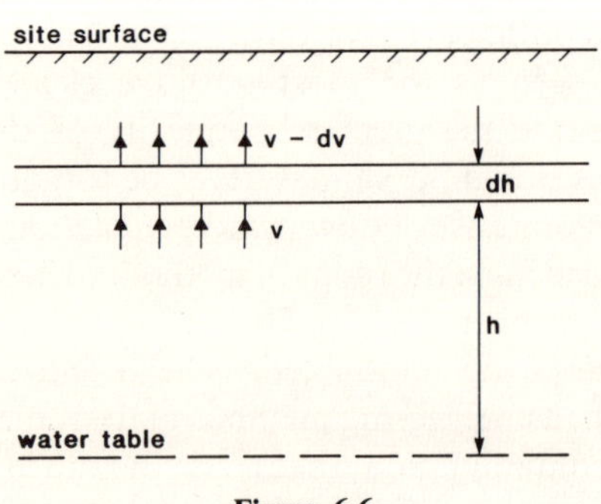

Figure 6.6

*By T. Sharrock, BEng, CEng, MICE, Acting Head of Building and Civil Engineering, Liverpool Polytechnic.

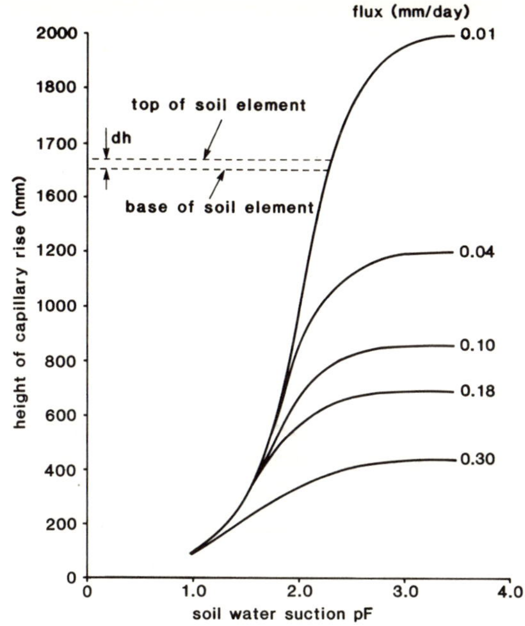

Figure 6.7

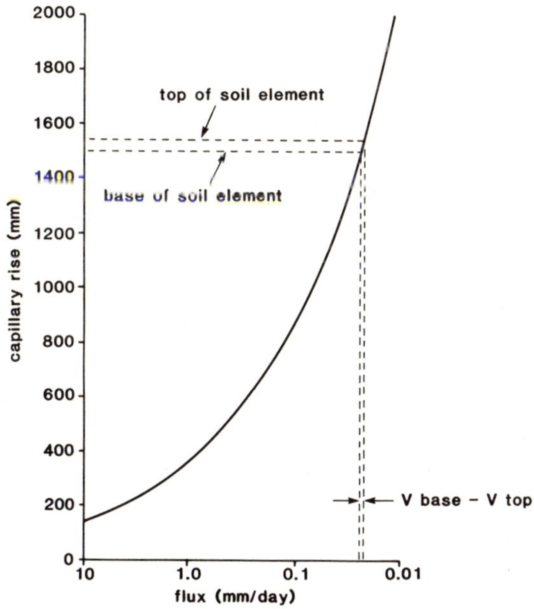

Figure 6.8

and the added concentration of the contaminant (in milligrams per kilogram of soil cover at that level) will be:

$$\frac{c \times \rho \times d(V\,\text{base} - V\,\text{top})}{\rho_{\text{soil}} \times \mathrm{d}h} \tag{6.4}$$

The terms c, ρ, and ρ_{soil} are of course obtained from the sub-surface exploration of the site and from the testing of the soil cover material. The value d is the design drought length and is obtained from meteorological data. The missing term:

$$\frac{V\,\text{base} - V\,\text{top}}{\mathrm{d}h}$$

is easily obtained by re-drawing Figure 6.7 to produce Figure 6.8.

An example of the use of Figure 6.8 is if the concentration of contaminant added to a 4 cm thick horizontal layer (lying between 154 cm and 150 cm above the groundwater table) is needed. The terms $c, \rho, \rho_{\text{soil}}$ in Equation (6.4) are already known, $\mathrm{d}h$ is 0.04 metres, and the differences in the flux just reaching the base and the top of the 4 cm thick layer can be read off the horizontal axis of Figure 6.8.

References

1. Wilson, D.C. and Stevens, C. (1981) Problems arising from the development of gas works and similar sites. AERE., Harwell, HMSO.
2. Naisbitt, D. (1982) Survey of prior industrial uses of N. Liverpool Sites. Unpublished research report, Liverpool Polytechnic.
3. Carson, R. (1963) *Silent Spring*. H. Hamilton (Pub.) [Hamish Hamilton (1983) Penguin, Harmondsworth (1982)]
4. Ames, S. (1979) Reclamation of land disturbed in mining *Proc. 3rd Ann. British Columbia Mine Reclamation Symposium*, Vernon, B.C., 311–324.
5. Figg, J.W., Sough, K.W. and Toblin, A.P. (1980) Some experiences of testing soils and materials from contaminated sites. In: *Proc. Conf. on Reclamation of Contaminated Land.* Society of the Chemical Industry, London Paper B5.
6. Parry, G.D.R. and Bell, R.M. (1985) Covering Systems. In: *Contaminated Land, Reclamation and Treatment.* M.A. Smith (ed.), Plenum Press, New York and London.
7. Interdepartmental Committee for the Redevelopment of Contaminated Land, (ICRCL) (1983) Guidance on the assessment and redevelopment of contaminated land. ICRCL., paper 59/83, HMSO., London.
8. Hillel, D. (1980) *Fundamentals of soil physics*, Academic Press.
9. Sharrock, T. (1986) Methods of evaluating soil cover materials and quantifying design proposals, *Civ. Eng. Tech.* (in press).
10. Bloemen, G.W. (1980) Calculation of steady state capillary rise from a groundwater table and through multi-layered soils. *Zeitschr. Pilanzenern*, **143**, 701–719.
11. McCarthy, M.J. (1980) Reclamation of a refuse tip for open space and housing development. In: *Proc. Conf. on Reclamation of Contaminated Land*, Society of the Chemical Industry, London, Paper B8.
12. Netherton, D.N. and Tollin, B. (1981) Reclamation of the Beckton Alps and adjacent areas. *Pub. Health Eng.*, **10**, (4), 202–221.
13. Croney, D., Coleman, J.D. and Bridge, P.M. (1952) The suction of moisture held in soils and other porous materials. DSIR Road Research Technical Paper 24. HMSO, London.

14. Hoeks, D. Institute of Land and Water Management, Wageningen, Holland—Personal communication.
15. Stieff, K. (1985) The long-term effectiveness of remedial measures. In: *Contaminated Land, Reclamation and Treatment*. M.A. Smith (ed.), Plenum Press, New York and London.
16. Cairney, T. (1982) In situ reclamation of contaminated land: the problem of a safe design. *Public Health Eng*., **10** (4), 215–218.
17. Institution of Civil Engineers, (1983) The influence of vegetation on the swelling and shrinking of clays. *Geotechnique*, **33** (2).
18. Lutton, R.J., Reagan, G.L. and Jones, L.E. Design and construction of cover for solid wastes land fill. U.S Environment Protection Agency, EPA 6000, 79–165.
19. US Bureau of Reclamation (1974) *Earth Manual*. US Gov. Printing Office, Washington, D.C.

7 Long-term monitoring of reclaimed sites

T. CAIRNEY

7.1 Introduction

Long term monitoring of contaminated land reclamation is important

(a) To give advance warning of any failure or the need for remedial action and
(b) To provide the data which can be used to evaluate earlier designs and allow cheaper and more robust solutions to be proposed in the future.

So far, such monitoring in Britain has been limited to a few research projects (1, 2) and to some sites (3) where methane gas production was a cause for concern. As is argued below, this lack of effort is perhaps the major limitation to a sensible evolution of demonstrably safe reclamations.

7.2 Established precedents for monitoring

7.2.1 *A comparison of the reclamation and civil engineering industries*

In some ways, the reclamation of contaminated land is another specialized activity in the wider civil engineering industry, comparable to the work of public health engineers and water engineers. If this association is correct, then a comparison of design attitudes is useful in emphasizing the technical benefits that monitoring and design evaluation can yield.

Civil engineering is, of course, a centuries' old activity that has gained from the services of many especially gifted individuals (of whom names such as Brunel and Stephenson are only those more obvious examples). Additionally it is a very large industry, attracting some 2000 graduate entrants in Britain each year, which has several hundred consulting and contracting firms and local government departments, and is supported by a range of governmental and industry-based research and information agencies (Building Research Establishment, Construction Industry Research and Information Association, etc.) and has ample publication outlets in the learned societies supported by several major professional bodies (e.g. the Institution of Civil Engineers, the Institution of Water Engineers and Scientists, etc.). Despite these advantages, civil engineering works do fail from time to time, and the industry, to improve its efficiency and profitability, now insists that each graduate entrant must undergo several years of continuing education and training before being recognized as professionally competent.

In contrast, the reclamation of chemically-contaminated land is a new activity in which few practitioners can claim more than a decade of experience. Very few organizations have so far been involved in reclamation work and the total technical population is probably only a few hundred people. No research and information agencies support the work, and no major professional body has yet to establish a learned society specifically for contaminated land reclamation. More crucially, no central registry of reclamation designs exists and the only process whereby design details are widely publicized are the few major conferences (4, 5, 6) that have so far been held.

Whilst the above comparison appears stark, it is not surprising in view of the short history of contaminated land activities, and the scale of land (see Chapter 1) that will sooner or later have to be reclaimed offers meaningful hope that as more and more organizations become involved, the overall experience will improve, and such limitations as the imprecision of the present contamination guidelines (see Chapter 4) will be removed.

However, the lack of long-term monitoring of reclamations, if this continues, will deny the opportunity to refine and improve reclamation designs.

7.2.2 *Monitoring, maintenance and 'factor of safety'*

The significance of monitoring is particularly clear in the civil engineering industry. There, near routine monitoring of new and innovate designs takes place. The foundations of structures, the settlement of various soil types when they are loaded, the stability of tall buildings and earth slopes, the efficiency of drainage works, the resistance of road surfacings and of other materials to service wear, and the maintenance needs of a whole range of works are monitored, often for very many years after construction.

This is done not only to protect the public who use these works from the consequences of any failures, but also to check the efficiencies of design assumptions and to allow cheaper and more robust structures to be designed on the proven performance of their predecessors. Every new advance in electronic sensing is pressed into such activity, and computer models, often of very great complexity, are used to mirror the effect of time and environmental factors.

It is thus not surprising that the civil engineering industry uses newly developed materials and construction techniques with routine confidence.

From this monitoring have come two concepts that are especially important to contaminated land activities—'design life' and 'factors of safety'.

To design a man-made work for an indefinitely long life is perhaps technically possible, but this would usually attract an unacceptable financial penalty and would entail the risk of catastrophic failure at some future time.

The British urban sewer system—to quote only a very simple example—was built largely in the late nineteenth century, without any particular note being taken of design life. The result of this is that urban sewers are now attracting public attention as they fail and collapse the overlying roads. The replacement sewers are designed usually for a proven 100 years of effective life, and in this case it would be technically possible to make the sewers last perhaps twice or three times as long, by making the sewer pipes stronger and bedding them in more robust foundations. However, the cost of this, in a time of public expenditure controls, would be unacceptable, and the concept of a specific design life is seen as preferable, in that it alerts the owners and operators to the need to replace obsolescent works before failure forces a more expensive and disruptive solution.

The same approach appears essential for contaminated land reclamations, particularly where these house buildings and structures and where peoples' health or lives might be at risk if failure occurs. However, as with the nineteenth century sewers, this is not yet possible, since monitoring data is lacking.

The concept of factors of safety is also important to contaminated land work. The idea of a factor of safety is often misunderstood and interpreted as meaning that, after each element in a design is sized to ensure that it is strong and robust enough for its duty, the designer then chooses to make it just that little stronger or more robust for the sake of caution. This viewpoint, however, misses the fundamental concept that a designer can never foresee the future, and that works might have to cope with environmental conditions far worse than any of our limited climatic records would suggest as likely to occur, or that some works might have demands put upon them that were never considered, or perhaps did not even exist, when the works were designed. In general terms, factors of safety have reduced in size over the years, as monitoring data has shown that some older works were over designed, and the result of this is cheaper designs that are more certainly safe.

Quite obviously factors of safety cannot be specified until performance data from earlier similar works is available, and so it is not possible to ascribe any precise safety levels to contaminated land reclamations.

7.3 Questions on the long-term effectiveness of reclamation options

The possibly adverse effects of time on various reclamation options have been briefly discussed in Chapter 5 and form the central theme of Stieff's (7) detailed consideration of the long term effectiveness of reclamation methods.

Of the methods listed in Chapter 5, only one—the total removal of all contaminated material from a site—is guaranteed to give long term safety. All other methods include some remnant risk, though current knowledge cannot quantify this.

Thus, for those on-site processing methods, which 'fix' rather than destroy contaminants, it can be argued that the stabilizing process need not be fully effective and could leave patches of still mobile contamination.

Equally, on-site methods which involve the even application of chemical or biological agents through the whole volume of contaminated wastes, might fail because of the inevitable variations in permeability, soil compaction and particle grain size which typify many contaminated sites. This is especially easy to imagine in cases where a simple application of lime is made to reduce the availability of metals for plant uptake. Apart from the difficulty in spreading the chemical agent evenly through the contaminated materials, the lime will gradually be washed out of the site by percolating rainfall, and will have to be replenished at some unknown frequency, if the site's pH is not to gradually fall.

Cover and encapsulation systems are even more suspect, since a range of physical, biological and chemical factors can attack the materials on which the reclamation's efficiency is based. A very simple example of such factors is if the seeds of deeper rooted trees are blown on to the site and allowed to root. In a few years, if the unwanted vegetation is not removed, the tree roots could either act as contaminant channel ways or even physically disrupt the soil cover, bentonite walls or concrete cut-offs on which the reclamations depend.

The hydraulic measures, used in some countries, to prevent leachate from contaminated sites polluting groundwater resources are especially suspect. Mechanical pumping plant wears out, the efficiencies of water abstraction wells decline as the wells silt up, and the complexities of groundwater flow paths can become even more complicated if local groundwater or adjacent surface water management is altered.

Thus whilst the longer term effectiveness of the various reclamation options cannot be factually quantified, enough doubt does exist to emphasize the need for long term monitoring.

7.4 Monitoring systems and methods

Once the case for monitoring is accepted, the difficulty is to decide

(a) What system of monitoring is required
(b) The frequency of the monitoring
(c) Its necessary duration.

These questions cannot be answered generally, since site specific factors—such as the persistence of the local contamination, and the sensitivity of the site re-use, must be considered.

Some contaminants are degradable and their effects will decay with time. Thus methane and the other gases can pose serious risks, but only for finite periods (see Chapter 11), and the pervasive phenol contamination on many old coal carbonization sites can be expected to reduce if the reclamation design includes site drainage works that allow water soluble contaminants to be washed to a central collecting and treatment facility. Other contaminants, however, are persistant and will continue to pose significant hazards worthy of long term monitoring. Combustible materials, if buried on a site, will continue to pose fire risks for the site's total re-use life.

The sensitivity of site re-use (Chapter 4, Table 4.2) is also important, since it is less easy to justify the expense and manpower for monitoring an amenity open area, where failure may result in the die-back of replaceable vegetation. On the other hand, a council estate with gardens carries with it the risk of large capital replacement costs and the possible harm to health and human life if critical failure occurs.

Thus the organization responsible for the reclamation should consider which of the following monitoring methods are justified.

7.4.1 *Periodic site inspections*

The cheapest and simplest monitoring consists of planned and systematic site inspections on an annual or seasonal frequency.

Visual observations are adequate to note areas of plant die-back, signs of erosion or of surface subsidence, the efficiency of surface drainage, and any migration of contaminated leachate into adjoining water courses or on to low points in the site. The value of such inspections is enhanced if a photographic record (taken from the same vantage points) is kept, and if an adequate system of retaining site inspection data is set up.

Such site inspections can be made more effective if facilities for measuring gas emission concentrations, groundwater levels, soil temperatures, etc. have been installed (see below).

7.4.2 *Periodic water quality analyses*

Equally simple is the routine sampling of drainage outfalls to monitor the changes in contamination levels. Such work is often undertaken by regional Water Authorities to protect water courses and to determine the loads imposed on their sewerage treatment works, and it is often possible to arrange Water Authority co-operation for a particular reclaimed site. If this is done, however, the Water Authority must be fully advised of the original contamination on the site to ensure that relevant analyses of drainage samples are carried out.

Such periodic water quality sampling (on say a quarterly basis) can be most useful in establishing the trends in contamination levels, and in identifying any migration of contaminants into adjoining land.

In some cases, where drainage works include the facilities to measure flow rates it is possible to enhance the simple quality data and to decide whether or not the drainage is behaving as designed and whether or not a fuller inspection is required.

7.4.3 *Periodic groundwater quality monitoring*

If groundwater quality is locally important, the Water Authority will normally insist that long term monitoring, outside of say an encapsulated site, is carried out. To do this requires specially designed observation boreholes, in which water level recording systems, and the facilities for water sampling (perhaps at various depths below the groundwater table) are installed.

Such boreholes are fairly expensive to establish and require specialist attention, thus such work is usually best undertaken in co-operation with the Water Authority.

7.4.4 *Routine site temperature recording*

In cases where deposits of combustible materials (coal, coke and pitch) have to be left on a site and are later covered with buildings and structures, a valid worry is that the combustible materials might be ignited by (say) buried power lines or if heated by fierce surface fires. Experience with such fires (8) indicates that they can be difficult to extinguish and may require massive excavation of the burning material and much of the surrounding heated soil. Obviously this would be highly expensive and disruptive for the site use postulated above.

Fortunately the temperature rise pattern associated with such events is quite diagnostic. Initially as the materials heat up, the soil temperature rises uniformly until the ignition point, when a much more rapid temperature rise occurs. The rate of change in the soil temperature is quite obvious and easily measured by a range of electronic thermocouple devices, available at a modest cost, and capable of being connected to data loggers, which can be pre-set to give automatic readings at a chosen frequency. Such devices can be buried, in manholes, adjacent to more important buildings and should give adequate warning of hazardous conditions.

Since the actual temperature at which ignition commences is variable and depends on the proportion of combustible matter in the soil, the soil's compaction, and on the ease or difficulty of air entry, through the cap layer that will normally be laid over combustible ground, it is thus a sensible policy to have pre-reclamation tests carried out to define the peak soil temperature that is hazardous and which requires remedial action to be initiated.

7.4.5 *Routine gas monitoring*

For sites with methane or other gas emissions, a monitoring of gas concentrations both in buildings and in the ground will be a normal precaution. Sensors capable of fulfilling this monitoring are available, and are suitable for both situations. These can also be connected to pre-set data loggers and can also activate warning systems.

Where gas concentrations in the site mass itself are required, specially constructed borehole probes are required (Chapter 11), and these can usually be used to monitor site groundwater levels and also to confirm the physical conditions of any soil cover laid over the site (see below).

7.4.6 *Site drainage efficiency monitoring*

Some of the cases outlined in Chapter 6 depended for their long-term safety on the adequacy of local groundwater control measures. In such cases, it is obvious that monitoring is essential, since the failure of drains by clogging is far from unknown.

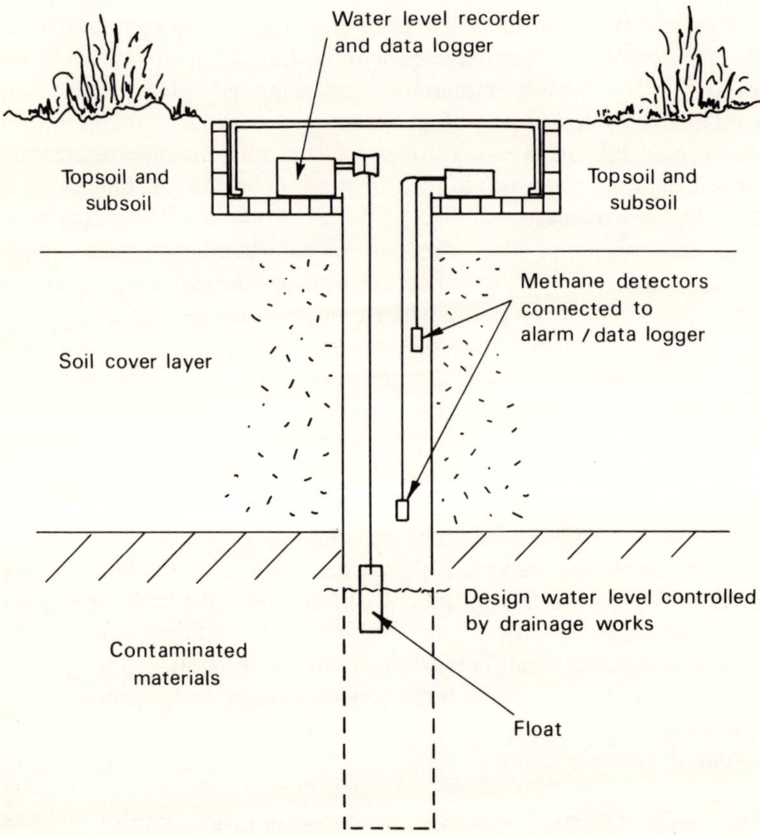

Figure 7.1 Simple water-level observation borehole also including facilities for methane detection

Thus a set of groundwater monitoring boreholes should be built over the site and each should have installed appropriate water level recording devices. Various recorders are commercially and cheaply available which can activate data-loggers (often via land lines from each borehole to a central data logger) and most can be adapted to set off alarm systems if the groundwater level rises above some specified level.

These boreholes (Figure 7.1) need be little deeper than the thickness of the soil cover over the site and can be used for gas monitoring and for confirming the long-term efficiency of the soil cover itself (see below).

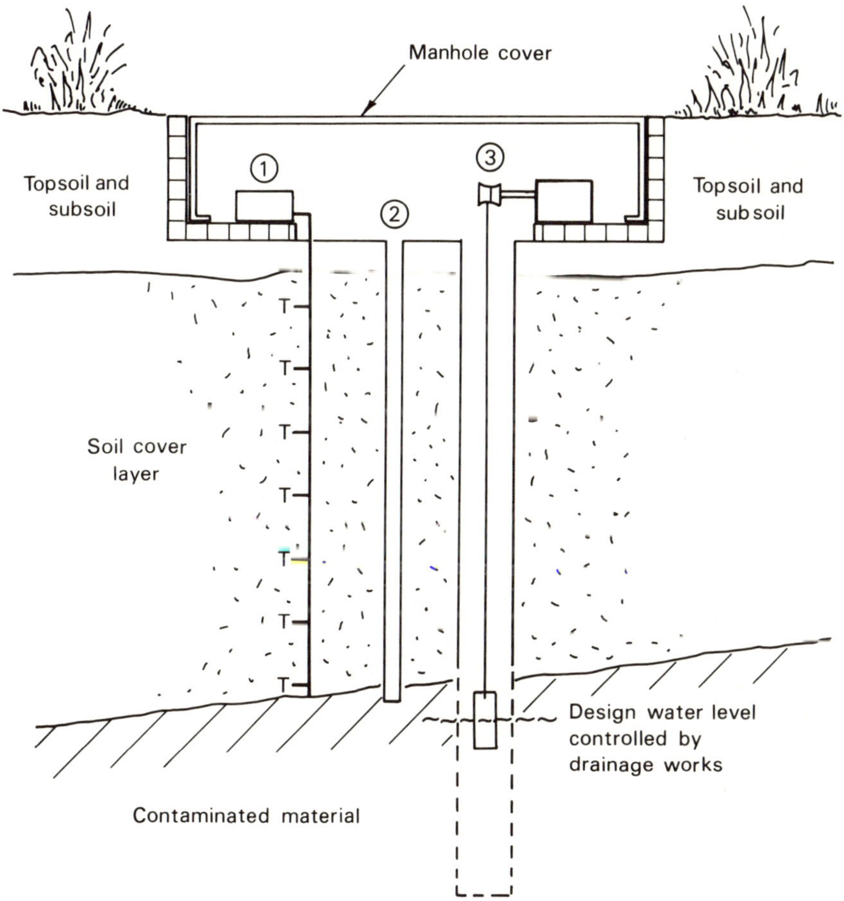

Figure 7.2 More complex monitoring system. (1), Tensiometer array connected to data logger to measure soil suction at points in the cover profile; (2), aluminium access tube for neutron probe, to obtain exact soil moisture values; (3), water level observation borehole with facilities for methane detection (if required).

7.4.7 *Soil cover monitoring*

A soil cover can fail if its physical properties change, particularly if the cover becomes silted up or if differential settlement in the underlying waste layers causes subsidence.

The physical property of the greatest concern is the cover's suction/water content relationship, and equipment is available to monitor this. An aluminium access tube, driven down to the design groundwater table allows a neutron probe to be used periodically to determine the actual moisture contents throughout the soil's vertical profile, and the suction values at chosen intervals down the profile can be measured by tensiometers connected to a suction electronic transducer and to a pre-set data logger (9). If the soil cover's suction/water content design properties are available from the laboratory process defined in Chapter 6, then the field monitoring data can be compared to these, and any deterioration of the soil cover's properties identified. Such deterioration will normally be seen as the soil cover having greater suctions (for any particular moisture content than that proven in the design process, and the effects of this can be evaluated by the Bloemen model (Chapter 6) and any necessary remedial action taken).

It would be reasonable to install several such suction/water content systems over the site's area and to include water level and (where necessary) gas concentration monitoring in the same systems. (Figure 7.2).

7.4.8 *Monitoring for specific contaminants*

It is now possible to detect the presence of particular chemical substances by specific electrodes, analogous to the measurement of acidity (pH) with a pH meter. Thus it is feasible to install specific electrodes below the surface of a site to give advance warning of undesirable conditions.

However, the long-term durability of such electrodes when buried in soil appears not to have been established, and any such monitoring would require specialist advice and might initially be of an experimental nature.

7.4.9 *General points*

The emphasis in the above brief notes is to utilize automatic data recording systems specially designed and installed before the reclaimed site is fully redeveloped.

This approach is seen as superior to ad-hoc later attempts at monitoring by excavating or drilling into the site, since access for such work will often be limited and could cause unnecessary alarm to site residents or users. It is also seen as desirable that monitoring systems are designed as part of the overall reclamation work and that these are chosen to minimise recurrent costs or manpower resources. The present efficiency of electronic sensing is such that this should not present serious difficulties.

7.5 Organization control of monitoring

Once the need for monitoring is accepted and a monitoring system is chosen to meet the needs of a particular reclaimed site, the remaining difficulty is to decide who should accept the liability and costs for the work.

As monitoring is only necessary to give advance warning of possible failure and to provide the data which will allow later reclamations to be more efficient, it is the author's view that local authorities or a central government agency should be responsible for such work. Private developers may well be conscientious in monitoring for initial periods, but it is unrealistic to assume that commercially biased organizations would be willing to undertake many years of routine measurement, particularly when any losses in reclamation efficiency are likely to be slow and gradual, rather than abrupt and dramatic.

Whether or not local authorities can accept yet further demands on their resources is uncertain, given the financial and manpower restrictions under which they operate, but without such co-operation the local authorities will continue to be troubled by contaminated land reclamation proposals, whose efficiency cannot be precisely evaluated.

The desirable ultimate situation is of course a central registry of reclamation data, available to local authorities and to commercial organizations.

7.6 Summary

Long-term monitoring of reclaimed sites has yet to become an established practice either in Britain or in any other developed country. Because of this, it has not been possible to give advice as specific as that in the other chapters of this book.

Despite the present situation, the need for long-term monitoring is obvious, both to protect the public and to improve the pool of technical competence.

The equipment necessary for monitoring is commercially available and usually at relatively modest costs, and the main difficulty appears to be the need to persuade local authority and governmental agencies to accept the responsibility for such work.

References

1. Jones, A.K. (1980) Monitoring of reclaimed contaminated sites. Report to the Department of the Environment (unpublished).
2. Welsh Office (1984) *Survey of contaminated land in Wales*. Welsh Office, Cardiff.

3. Peckson, G.N. (1985) Methane and the development of derelict land. London Environmental Supplement, 13.
4. Society of Chemical Industry (1980) *Proc. Conf. Reclamation of Contaminated Land.* Eastbourne 1979, SCI, London.
5. Greater London Council (1984) Contaminated land—the London experience. London Environmental Supplement, 7.
6. Assink, J.W. and van den Bruik, J.W. (1986) Contaminated soil. In: *Proc. 1st Int. Conf. on Contaminated Soil*, Utrecht, 1985, Martinus Nijhoff, Dordrecht.
7. Stieff, K. (1985) The long term effectiveness of remedial measures. In: *Contaminated Land Reclamation and Treatment*, M.A. Smith (ed.), Plenum, New York and London.
8. Beever, P.F. (1986) Assessment of fire hazards in contaminated land. In: *Proc. 1st Int. Conf. on Contaminated Soil*, Utrecht, 1985, Martinus Nijhoff, Dordrecht, 515–522.
9. Blackwell, P.S. and Elsworth, M.J. (1980) A system for automatically measuring and recording soil water potential and rainfall. *Agric. Water Management* **3**, 135–141.

8 Safety in site reclamation

D.L. BARRY

The range of hazards that can be present during the redevelopment of contaminated sites can be very wide, and measures to protect workers engaged either in investigation or construction activities must match the risks involved. Detailed specialist advice from a suitably experienced occupational hygienist is essential when, for example, volatile organics and carcinogenic dusts are likely to be encountered. However, not all sites labelled as 'contaminated' are likely to be highly hazardous and may have, at worst, undesirable levels of contamination with respect to the proposed re-use or to the health of site workers. Such sites include those that have been largely 'cleaned up' as a specialist operation in advance of the main redevelopment contract works, and also sites that have relatively low levels of contamination in the first instance. In either case the risk of encountering critical chemical hazards is frequently low, but not always so low that it can be ignored.

The context of advice and discussion in this chapter is where conventional engineering works are being carried out and where the majority of the workforce is generally unfamiliar with chemically hazardous conditions. Therefore, the emphasis is placed on the provision of (i) adequate safety facilities, and (ii) an effective set of safety procedures that help avoid a critical outcome from a hazardous occurrence, without unduly interfering with conventional construction techniques.

8.1 Introduction

In the process of redeveloping contaminated sites, the hazards to construction workers are frequently overlooked. This is ironic since, in many instances, not only are they the most exposed 'target' group but they may be involved in the installation of measures for the protection of future users. As with many aspects of redevelopment, the recognition of safety problems and the adoption of relevant pragmatic procedures are not synonymous. All too often insufficient attention is paid to the problem during the design stage, and removing contaminated materials, for example, is sometimes seen simply as a conventional earthworks operation.

It is incumbent upon designers and developers to set out within the contract documents for construction works the need for appropriate measures or procedures to safeguard worker health. It is not sufficient to hide behind the statutory requirements of the Health and Safety at Work etc. Act 1974, (1) which requires the contractor to safeguard his workforce. Moreover, apart from any moral obligations, there can be enlightened self-interest in trying to ensure that contractors, when tendering for relevant works, are fully aware of the employer's intentions vis à vis safety strategy. Contractors who are awarded contracts and then find that there are unexpected stringent and perhaps costly safety requirements can be troublesome. Had such measures and procedures been included in the tender, not only would the tendering exercise be more equitable for all tenderers, but worker safety would be less likely to be compromised.

Many guidance and technical notes are issued which relate to safety aspects relevant to contaminated sites (1–5), and the reader is recommended to consult these. In addition, papers have been presented at a number of conferences on contaminated land, which have been referred to in other chapters (e.g. the IVth International Environment and Safety Conference, Barbican Centre, London, March 27–29, 1984).

8.2 Nature of hazards

8.2.1 *Introduction*

Hazards can be present in three principal forms, namely solid, liquid or gaseous, and can occur in a wide range of circumstances. They can be presented by flammable or explosive substances or through the possibility of inhalation, ingestion or physical contact with chemical compounds. The risk associated with such interactions is fundamentally related to the physical context of the works, the particular chemical compounds and their form, and the degree of exposure. The physiological status of the worker is also of relevance. Thus inhalation of asbestos fibres or splashing of chemicals in the eyes can be very serious for different reasons.

The large majority of hazards are likely to be encountered during excavations for foundations or drainage systems or during the regrading of the ground mass. In the first two of these operations, manual workers frequently either come into contact with the contaminated ground or are in a confined, unventilated and potentially aggressive environment. It must also be kept in mind that hazards may have an impact on the site neighbourhood, e.g. through fugitive dust or vapour migration. Although this is of low probability on the sites under consideration, chemical hazards once encountered must be viewed in a wider context than the immediate vicinity.

8.2.2 *Solids*

Hazardous chemical compounds can occur in many physical forms (e.g. powdery, granular or crystalline) and can also be incorporated within the matrix of otherwise 'inert' materials found on most sites. The greatest risk is usually in respect of any respirable compounds which become fugitive when disturbed. That, however, is based on the reasonable assumption that site workers do not ingest such chemicals except perhaps through handling foods with soiled hands or by hand-to-mouth contact (through smoking for example). Excessive inhalation of dusts can at best be unpleasant and cause discomfort, but at worst could prove fatal or cause serious pulmonary disorders, asbestos being a particular example (see 8.2.4 below). Asbestosis, a debilitating fibrosis of the lung, is associated with chronic exposure. However, the development of asbestos-associated carcinomas may result from far shorter exposure times.

The visual or olfactory evidence of potentially harmful substances can often be quite obvious, e.g. they may have 'unusual' colours (such as 'blue billy' on gas works sites) or odours, but they might also be masked by seemingly innocuous, less harmful, or naturally occurring materials. On balance, however, if the potentially hazardous materials are thus masked, they are less often likely to become fugitive when disturbed. Therefore, in practical terms, such hazards might present a low risk. The exception to this could be liquids that are absorbed by the host solid. Thus phenols, for example, might not be readily perceived within the soil matrix and yet may still induce dermatitis.

8.2.3 *Liquids*

Many of the hazardous liquids likely to be encountered are also likely to have either unusual odours or colours and so can present an 'early warning' to workers. Liquids can be hazardous in a number of ways; they can be corrosive to the skin, they can release toxic or inflammable vapours when exposed to the air, they can cause eye-damage through splashing, or if consumed accidentally, (if a worker falls into them, for example), they can be most dangerous. When drums or tanks or other containers of liquids are encountered, expert advice must be sought before any 'remedial' action is considered since hazardous or explosive fumes can result, for example, if the contents are suddenly exposed to air. Labels on such containers cannot be relied upon to define the contents.

Liquids can be physically more difficult to deal with than solids (they cannot always be 'dug out' with a mechanical digger and placed in an open tank, for instance) and they must be treated with even greater caution. Liquids entering trenches are common, particularly when old drains and pipes are broken out, and the resulting problems relate not only to the immediate health hazard (e.g. splashing or fumes) but also the difficulty of clearing the contaminated trench to an acceptable level.

It is not appropriate to give generalized advice for dealing with liquids because of the wide range of possible risks associated with them. For example, it is probably an instinctive reaction to try to prevent the spread of a spilled liquid by removing the container to a 'safer' location. Such action may at best be inappropriate and, at worst, very hazardous; for example, the container may be in an unstable condition due to corrosion. Similarly, if a digger driver (in a suitably ventilated cab) works from the upwind side of a spillage, then, unless the liquids are flammable or explosive, some action such as covering with clean soil might seem appropriate in the first instance. That, however, can create a greater mass of affected material to be dealt with subsequently. The best procedure, therefore, is to vacate the immediate area and inform the Environmental Safety Officer (see 8.4.3) or similarly qualified person. (The scenarios above assume encounters with either unknown liquids or those which, based on the site's history, could be present and which are known to be hazardous.)

8.2.4 *Gases*

Toxic and asphyxiant gases can occur even in natural ground, e.g. acid ground conditions and carbonaceous material can create carbon dioxide. Similarly, oxygen deficiency can occur in sulphide-rich clays, and methane can be present in peaty or silty conditions. It is not unreasonable, therefore, to expect that on contaminated sites there will be an even greater likelihood of toxic, asphyxiant or flammable gases. The presence of these gases is most likely to be critical in confined spaces where little or no ventilation occurs. Thus, trenches or deep excavations warrant particular attention. Also, gases that are heavier than air can accumulate at the excavation base where accessibility/aggressibility is often difficult. The hazards in such circumstances are greatly enhanced when the gases exhibit no odour and are invisible, thus offering no warning of their presence. Leaving trenches, shafts or manholes over weekends, for example, can increase the possibility of hazardous gas accumulations, especially since air movements brought about through normal working activities have ceased. Gas tests should always be carried out in such circumstances before the spaces are entered on resumption of work.

The presence of detectable concentrations of gases does not mean that a hazard exists. Occupational Exposure Levels (OELs) have been established for an extensive range of gases and these can be consulted (2) for guidance on safe levels. In practical terms, however, any hazardous gases detected must be treated with caution since the location at which the concentrations were measured might not represent the worst situation.

For odorous gases the sensory system can obviously be very important in the first instance, particularly where the odour threshold is considerably less than the OEL. Account must be taken of sensory loss which is a particular problem with hydrogen sulphide (H_2S)—it has an odour threshold of 0.0025 ppm, but at high concentrations causes fatigue of the sensory system, thus giving the impression that the gas is no longer present.

8.2.5 *Asbestos*

The threat presented by airborne asbestos is an ever-increasing problem on many sites, and its unpredictable presence can make it an even more difficult hazard to counter. The health risks from asbestos have been discussed extensively in the technical and other media, and standards have been introduced for the maximum permissible levels in the worker environment (3). Indeed, the maximum permissible values would appear to be based as much on the facility by which they can be measured 'in the field' as on the scale of hazard that they represent.

The most critical fibres are considered to be those of 'medium' size. This is because the very 'small' fibres do not come into contact with the lining to the lung if inhaled, while the 'large' ones are likely to be trapped by the natural filtering mechanism and exhaled or coughed out. Thus it is the fibres that adhere to the lung that are of most concern. However, in practice it is not meaningful to try to distinguish between fibre sizes, and so all asbestos must be considered equally hazardous. This generality also applies to the three main types, namely white, brown or blue (chrysotile, amosite and crocidolite respectively), although clinical evidence appears to suggest a slightly smaller risk from the white form.

Any encounter with and proposed remedial action relating to asbestos must take full account of the relevant statutory procedures (4). However, if loose asbestos is encountered it is important to spray it carefully with water as soon as possible to prevent its migration. This ensures that both workers and the neighbourhood are protected while a remedial strategy is being devised. Subsequent control and prevention requirements must be based on the particular circumstances prevailing. Since many local authorities are now more accustomed to dealing with this problem than with chemical contamination, a greater pool of knowledge is more readily available.

8.3 Protection against hazards

8.3.1 *Introduction*

In determining the most appropriate protection against the various types of hazards discussed earlier, it is necessary to take account of the construction methods likely to be employed on the works. For example, the effect of bored piles is fundamentally different from that of driven piles since the latter do not involve the removal of any existing ground mass. Also, because many site workers may be relatively unused to these hazards, it is important that they are not required to climb too many 'learning curves' simultaneously by (i) dramatically changing normal working procedures or techniques and (ii) introducing unnecessary additional protective facilities that can actually hamper their normal efficiency and comfort, and hence their safety. This does not imply that personal safety should be compromised, but merely that pragmatism must be a critical factor in forming an effective safety policy. Indeed, over-cautious procedures and measures can bring safety policies into disrepute, particularly if their relevance becomes less obvious when no hazards are encountered as works proceed. Protection for workers can be provided in the form of clothing, hygiene and safety equipment, for example, and does not necessarily mean sophisticated or expensive items. Each site must be assessed in its own right, and proper account taken of known historical and chemical data in determining the nature and scale of necessary precautions, whilst always recognizing the potential for encountering unexpected hazards.

8.3.2 *Clothing*

The simplest and most general forms of protective clothing are overalls, boots and gloves. For a large majority of sites such measures are likely to be adequate. However, the nature of these items may have to be amended from time to time to reflect their robustness or durability in the face of specific conditions. Thus, in line with the general philosophy outlined above, it is essential to recognize the physical conditions under which engineering works are usually carried out; for example, harsh weather is a factor which does not always cause works to be halted. Accordingly, the durability of protective clothing and its suitability in extremes of climate must be clearly recognized. Disposable 'paper-like' suits are very useful when dealing with occasional dirty jobs but are not suitable as a basic engineering outfit.

Overalls or boiler suits should ideally be made of Terylene or similar materials, but in the reasonable interests of worker comfort, polycotton is normally allowable. In both cases, however, it is important that they have close-fitting or elasticated ankles and wrists. Overalls should be laundered regularly, and if they become soiled with asbestos, tars, oils, solvents and similar liquids, they should be changed and not re-used before specialist laundering. It is also advisable that overalls are not taken off-site at the end of work periods, but are left in the site changing room. The responsibility for suitable laundering must lie with the employer, but the worker must also share responsibility in terms of observing the need for it.

Steel-protected boots of the wellington-type are usually preferable, but if liquid organic compounds are prevalent, ordinary rubber boots will not retain their protective integrity. Accordingly, a suitable alternative product must be chosen or, at the very least, the normal boots should be inspected frequently for signs of deterioration. PVC-coated gloves are likely to offer suitable protection in a large number of instances but, as with rubber and plastic products, they are not resistant to some organic compounds. Gloves coated with PVA (polyvinyl alcohol) for example can offer protection against organics, but they are not robust and are permeable to water. Therefore, if extensive contact with organic or similarly aggressive compounds is likely, it is important to examine closely the proposed manner of working, so that such critical interaction is avoided or reduced to a minimum. This can often be achieved by increasing the use of mechanical plant, for example, or by modifying the construction process in that location.

The possible need to review working methods and/or safety procedures highlights the importance of allowing a suitable contingency, of both money and time, in the works contract.

8.3.3 *Personal hygiene*

A further basic precaution against chemical hazards concerns personal hygiene. In this regard, washing hands before eating and avoiding hand-to-mouth contact as far as possible can achieve an effective reduction in risk. There are, however, two main prerequisites to the former; one, suitable washing facilities must be provided, and two, washing must be carried out effectively. Effectiveness of washing can be greatly enhanced by the use of a suitable hand barrier-cream which should be applied (to clean hands) before every work period. Such creams can offer some protection against casual interaction with contaminants and also help in protection of cuts and abrasions. Cream is, however, not a substitute for gloves. Equally, gloves do not obviate the need for cream, since gloves are often removed to improve dexterity for some tasks.

Showers are now a common facility on construction sites and are usually essential on contaminated sites. A drench shower is also desirable for effective and rapid dousing when necessary. Eating facilities also are normally provided on construction sites and these should be located so that contaminated dusts are unlikely to enter. Boots and protective clothing should not be brought into such facilities if critically soiled, since contaminated mud can dry out, thus creating toxic dusts in the internal environment.

A general ban on smoking on all parts of contaminated sites is not always considered necessary, although parts of each site (such as trenches) should have such absolute prohibitions. However, notwithstanding the risk of causing an explosion, smoking does increase the risk of ingesting contaminated soil through increased hand-to-mouth contact. (On grossly contaminated sites with airborne contamination, chemical toxicity can be enhanced when the contamination is drawn through a lighted cigarette.)

8.3.4 *Hazard monitoring*

Monitoring for chemical hazards takes two principal forms; one, the visual and olfactory senses and two, the use of detection devices. The former should be carried out not only by the Environmental Safety Officer (see 8.4.3) but by all site workers. In many cases, this can be the most effective warning method but one which cannot be relied upon totally. Accordingly, some detection devices are needed, and these invariably relate to gases. Solids and liquids would, of course, be sampled and analysed on an *ad-hoc* basis, but potential hazards in excavations, in particular, must be monitored in a regular, comprehensive manner.

Gas monitoring is usually carried out either by automatic gas detectors or, for example, by 'Draeger' tubes. In either case, specific or generic gas types must first be listed by taking account of site history, the conditions observed or analysed and any relevant experience on other contaminated sites. Notwithstanding this, flammability and oxygen concentration measurements are appropriate on all sites. Hydrogen sulphide or organic vapours are frequently on a short-list of gases that require monitoring. Instantaneous, or at least rapid, analysis is needed if the system is to be effective in practice.

A practical approach is to select gases that can act as indicators (directly or indirectly) for other gases; for example, carbon dioxide in critical concentrations would be indicated indirectly by a deficiency in oxygen and so, normally, need not be monitored.

Automatic gas detectors with multiple gas sensors and adequate weather protection can be suitable for use on construction sites. Preferably, these should have clear audible alarms and should be set into excavations in a zone which is most likely to experience a source of hazard. Monitoring should be carried out before workers enter confined spaces, a policy that should apply to all construction sites, particularly where deep trenches or sewers are involved. Sewers can act as carriers for critical gases even before being connected to the main sewer system.

Testing for gases in advance of trenching can be beneficial, particularly in suspect areas such as where oil spillages and the like may have taken place. Anticipating hazards is not only a good safety policy but also can bring considerable practical benefits if delays due to 'unexpected' conditions can be avoided.

8.3.5 *Safety equipment*

A list of those reasonable items that should be provided on the sites under consideration is as follows:

(i) Face masks and filters
(ii) Breathing apparatus
(iii) Safety harness and lanyards
(iv) Fire extinguishers
(v) Drench shower
(vi) First aid (see 8.3.6)

The suppliers of safety equipment in general are listed in such publications as *Safetymate* (6).

Breathing apparatus. Broadly, this can be of two types, self-contained and remote. Self-contained equipment uses compressed air supplied from a back cylinder to a full face mask, while the remote type consists of a face mask connected to a length of hosepipe through which air is drawn from a 'remote' location. The latter supply can be compressed air cylinder or 'fresh' air. Both of these types have their advantages and disadvantages; for example, the former enables a rescuer or worker to operate 'independently' of others, but the weight and bulk of such items can sometimes be cumbersome; the latter has the advantage of being easy to don, is light and is reasonably maintenance-free, but the hosepipe length can be restrictive (in a deep trench for example). The positive pressure of self-contained systems offers more protection, as the excess air from the supply cylinder prevents toxic gases entering the face mask. It is appropriate to have both types on site, together with a special lightweight full-face respirator for a victim being rescued. 'Escape' sets, which are small self-contained systems, could also be carried to a victim.

Safety harness and lanyards. These are needed when entering a trench to effect a rescue or when a hazardous gas environment is suspected. (In the latter cases, suitable breathing apparatus is also essential.) The harness should be a secure full-body type and must be securely tethered with a lanyard to a fixed object at the surface to prevent it falling into the trench. At least two workers are needed to haul a person from a trench. A harness and lanyard are also needed for attaching to the person being rescued. Although less effective than a full harness, an industrial belt is a practical feature since (i) it can be worn permanently by all trench workers with little or no discomfort, and (ii) it can then be readily attached to a lanyard in the case of a collapsed trench worker (Full harnesses must be 'stepped' into.) It must be recognized that an industrial belt, if used on its own for a rescue, could cause or worsen physical injuries, but speed of rescue from a hazardous environment might be more critical in some circumstances. Harnesses should be simple enough to be donned quickly in an emergency.

Face masks. Face masks must not be used when oxygen levels are depressed. Ori-nasal masks are basically for filtering gases or dusts from the air and can consist of a 'simple' dust mask such as worn by operatives cutting concrete, or a respirator with one or two special filters. Filters must be chosen to match a specific hazard and not used as a 'general precaution'. Masks can be uncomfortable to wear for protracted periods, and restrict speech. Therefore, consideration should be given to providing forced ventilation of the working area when the hazard or discomfort persists (see 8.4.7).

8.3.6 *First aid*

The content and size of first-aid boxes for industrial sites has been set out in HSE (7), the size of workforce being an important factor. A stretcher and blankets, for example, are also necessary, as are eye-wash kits, the quantities of liquids for which can be easily underestimated. An experienced 'first-aider' is essential on any site, quite apart from the risks due to ground contamination.

8.3.7 *Equipment storage and maintenance*

As with any sensitive, critical or vulnerable item, it is essential that safety and monitoring equipment are properly stored and maintained. Since vandalism is a major current problem, not least on construction sites, a secure steel cabin is probably needed for equipment storage. This should be suitably equipped with shelving and working space and be kept dry. Also, its siting is important in terms of ease of access. A range of items should also be stored in a suitable transportable 'trunk' to be kept close by excavations and other critical areas. The contents should include breathing apparatus, safety harnesses and lanyards, fire extinguisher, first aid kit, blanket and 'klaxon' (see 8.4.8). Simple instructions can be attached to the lid of the trunk to remind rescuers of the correct procedures to be followed.

Maintenance should accord with manufacturers' instructions. Particular care should be taken with expendable items such as gas cylinders, filters, and the like so that stocks are not found to be critically low in an emergency. Equipment that is not properly maintained could be more dangerous than no equipment, since a false sense of security can be gained from the former.

8.4 Safety procedures

8.4.1 *Introduction*

In specifying or controlling working procedures a balance must be drawn between safety, reasonableness and practicability. Also, the principles and details of the safety procedures should be included in the contract documents for the engineering works. There are two principal reasons for this: (i) safety controls and responsibilities must be clearly defined in any contract, and (ii) by specifying the minimum acceptable procedures and equipment, the tendering process is more equitable (i.e. the more safety-conscious contractor is not penalized by pricing properly for relevant measures). Moreover, it is essential that all tenderers for the works are made aware of the risks of encountering chemical hazards on the site by making available site investigation reports.

8.4.2 *Contract powers*

The following is an example of how powers to enforce safety policy could be defined in a works contract:

> The Council shall have the authority to inspect any part of the Works during the performance of the Contract and to approve or condemn the method of working. The Council shall have the power to suspend any part of the Works if the Council consider that the presence of certain materials or the Contractor's method of working represents an unacceptable risk to site workers or to the general public. Such suspension shall remain in force until the Council is satisfied that work can proceed safely.

While such powers can be easily written into a contract, it is important that they are not abused by an over-cautious or indecisive approach in practice. This could lead to an undermining of the value of safety principles and could also present significant contractual and financial problems. Again, if the hazards are extensive and the risks of encountering them are high, then consideration should be given to using a specialist contractor to remove relevant materials before the redevelopment works commence. Moreover, if such hazards are found in the event to be greater than acceptable, consideration may have to be given to engaging specialist contractors to deal with the problems.

8.4.3 *Environmental safety officer*

The need for a full-time Safety Officer will depend on the scale and nature of risks involved. Defining such a need is not simple, but account must be taken of (i) the known or suspected hazards, (ii) the risks of such hazards arising, and (iii) the duration of the works or periods during which the hazards are most likely to occur. The likelihood of dealing with further similar redevelopment projects may also influence the appointment (temporary or otherwise) of a Safety Officer. In the case of some engineering contractors, such an appointment is more likely than not, while a local authority or private developer may have considerable difficulty in making a short-term appointment. Consultants are usually engaged to overcome such problems.

Where appointed, a Safety Officer should police all chemical safety matters on site and ensure that safety principles are not violated. The Safety Officer should preferably have a chemistry background with considerable experience in gas monitoring and sampling and, if engaged by the 'works employer', should form part of the resident site staff under the Resident Engineer. His background experience should allow him to respond effectively to visual and olfactory evidence as well as interpreting measured data. Matters relating to construction safety might not be an immediate part of his remit, but to aspire to this would be most desirable, e.g. trench stability is a significant hazard on many sites and particularly so on contaminated sites where the ground has usually been disturbed by previous activities.

The presence of an employer's Safety Officer does not and cannot reduce the Contractor's responsibility for the health and safety of his workforce; this responsibility is laid down in the Health and Safety at Work Etc. Act 1974 (1). It must be remembered that all persons on the site share some of the responsibility for their own safety. ('Employer' here means the person or persons for whom the works are being carried out.)

8.4.4 *Site training*

It might be said that the risk of accidents from obvious hazards is less than that from hidden or invisible hazards. Accordingly, a key element in any safety policy is in ensuring that site workers in general, and particularly those most likely to encounter hazards, are made aware of the types of hazards and degrees of risks associated with them. While construction workers cannot be expected, except through experience, to recognize many of the likely hazards on a contaminated site, their reasonable alertness to 'unusual' circumstances is often sufficient to prevent problems arising. Therefore at the outset of a contract, and as frequently thereafter as circumstances suggest, all site staff, and not just the labourers, should be addressed. Obviously the co-operation of the contractor's site manager is needed (bearing in mind that as far as his employees are concerned, such address or similar instruction, is his responsibility under the law).

The safety policy and procedures should be clearly outlined to site staff and the types of safety and monitoring equipment demonstrated. Simple hand-outs, such as that presented in section 8.6, should be provided and effective display boards or posters erected.

8.4.5 *Safe working procedures*

A simple 'permit to work' system may need to be initiated from the outset of the works. These should be related to specific areas on the site known as 'Restricted Areas'. Only those workers with valid permits should be allowed to enter or work in these areas, which can be marked off with fluorescent tape, for example. The permit system could be operated by issuing a 'red helmet', for instance, to relevant workers who would, in turn, be required to surrender to the Safety Officer their matches and cigarette lighters. Having a different coloured helmet is a practical permit system since (i) it is clearly visible and (ii) helmets must be worn in any event. (At the end of each working day, helmets can be re-exchanged for smoking materials.) Restricted Areas should be defined to correspond with, for example, excavations or any other feature considered by the Safety Officer to present an unacceptable risk. Entry to the Restricted Areas should also be contingent upon the wearing of suitable protective clothing.

All workers should wear suitable overalls and wellington/safety boots at all times and, depending on the actual or probable conditions, gloves and safety spectacles/goggles. The wearing of gloves at all times need not be mandatory, especially if dexterity is unnecessarily reduced as a consequence. Similarly the wearing of safety spectacles or goggles must be viewed pragmatically, since misting up and restricted peripheral vision can result.

8.4.6 *Excavation works in restricted areas*

The following are examples of conditions that should relate to working in excavations which are designated as Restricted Areas:

(i) No one to work in Restricted Area without permit
(ii) No one to work alone in Restricted Area
(iii) No one to enter an excavation greater than 1.5 m in depth until the excavation has been gas-tested and cleared by the Safety Officer
(iv) A ladder must be placed in the trench at a convenient location
(v) No one to enter an excavation unless the following items are immediately to hand; self-contained breathing apparatus, safety harness plus lanyard (one each for rescuer and victim), powder fire extinguisher, and audible alarm device
(vi) At least three persons must be present when working in excavations, at least one of whom must always remain at the surface.

The question of the number of people needed at a trench excavation is not simply answered. Without doubt at least two people are needed to haul one person from a trench, but the philosophy being expressed here is that the person 'on duty' at the top of the trench must not enter a trench and so cut off the option for summoning further help.

Since hazardous gaseous conditions are most likely to occur near the face of an excavation, it may be prudent to instal a continuous gas monitor in that zone. If the method of trench support used by the contractor involves a steel shield (as opposed to the more traditional trench shoring using steel sheeting and struts), the gas detector unit can be hung from the front of the shield so that it can detect gases that could migrate into the working zone. No worker should, in any circumstances, work in advance of the shield, thus ensuring a primary, albeit insufficient, precaution. Since any gas encountered may migrate along the trench and not merely rise, the greater the distance from the source to the workers the greater the opportunity for the gas to become diluted.

The gases selected for such permanent monitoring would normally be oxygen, flammables, and such toxic gases (e.g. H_2S) as are most likely to occur, based on the known history of the site. Particular attention must also be paid to observations by trench workers of the environment in the trench. The simple message for all trench workers must be that if they feel drowsy, sick or 'unusual', to *get out,* and also to tell their workmates in the trench.

It must be stressed to workers that they should be particularly careful with old pipes or land drains encountered in excavations, since these can contain hazardous liquids. Not only could the careless breaking of such pipes create a hazard, but the liquid once released into the trench could create unacceptable vapour concentrations. Moreover, the cost of clearing out the consequent contamination could be high in both time and money. Where encountered, therefore, such pipes should be carefully examined before being broken into. The liquids should be collected and the exposed or broken pipe-ends should be stopped up effectively. Backfilling to old trenches, being more permeable than the host ground, can also be the source of chemical liquids entering an excavation.

8.4.7 *Abatement or removal of hazards*

When a hazard is encountered and prevails, some measures must be adopted to allow work to continue safely or protect the neighbourhood. Almost invariably, a chemical hazard that is not removed will remain a hazard, and so it is advisable, if not essential, to remove it even if it does not curtail the operations in hand. Since the scale of an 'unexpected' hazard cannot be predicted, advice on the most effective manner for dealing with it cannot be given in advance. Therefore, as for grossly contaminated sites, specialist advice must be sought, particularly when drums or volatile organics are encountered.

There are, however, many circumstances where the scale of hazard is modest, and within the scope of readily available facilities on site. Excavating contaminated ground can be straightforward if access for mechanical plant is reasonable and ground stability is not a problem.

Dealing with asbestos requires strict adherence to the statutory requirements with respect to disposal (4). The removal operation is, strictly speaking, not controlled directly by legislation, but the available codes of practice should be followed. In simple terms, wearing disposable suits, gloves, boots and appropriate respirators, together with adequate water spraying, should enable asbestos to be at least removed from the particular location where it created a work impediment.

Vapours in trenches cannot always be readily countered since the source may be remote or unknown, e.g. gas migration through old pipes or sewer lines, or even through new sewer lines. Moreover, it may be impractical and even unnecessary to act, if the conditions do not constitute a hazard of any consequence in the future. Thus the most effective abatement measure is to introduce a level of ventilation that will suitably dilute the concentrations to desirable values. Ventilation is preferred to breathing apparatus, since the latter can be cumbersome and even impractical in some circumstances. The ventilation system should obviously be operated so as to take the gas away from the working zone, by blowing or sucking. In choosing a ventilation rate and a 'suck or blow' option, account should be taken of the effects on workers of excessive draughts, particularly in cold weather.

8.4.8 *Emergency procedures*

A number of very important preparations must be made before works start so that the scale of any emergency is reduced, or even prevented. These preparations should include the following:

(i) Inform the local emergency services that works are being carried out, indicate the nature of possible hazards and advise on site location and access
(ii) Establish critical telephone numbers of emergency services
(iii) Arrange easy access to a telephone (on site, if possible)
(iv) Clearly instruct workers on emergency procedures (see Appendix)
(v) Ensure that vehicular access to the site is clear.

The example given in the Appendix (8.6.2) on 'emergency procedures' relates to operations in a 'Restricted Area', but the principles are applicable to the whole site. Also, while the example deals principally with trenches, the essence of the advice can be applied throughout the site.

Apart from emergencies created through 'normal' site operations (such as trench collapses), the most likely emergencies are likely to relate to (i) highly flammable or corrosive liquids, or (ii) worker collapse in an excavation due to hazardous gases.

Sounding the alarm A compressed-air hand-operated klaxon stored in the transportable safety 'trunk' is likely to be appropriate on most sites, with a back-up alarm housed in an item of mobile plant. Since virtually all sites have mechanical diggers, and because such diggers are very frequently at or close to excavations, for example, they are appropriate for housing an alarm. Alarms should have a pre-arranged pattern.

Responding to alarm. The nature of the emergency should be communicated as soon as possible to the appropriate services. Suitable first-aid procedures should be followed and every effort made to prevent an escalation of the problem by, for example, other workers being put seriously at risk.

8.5 Conclusions

It is not always easy to define the level of precautions that should be taken in the face of a particular set of hazards, especially when the risk of those hazards occurring is not accurately predictable. As with many problems which cannot be judged on a purely scientific basis, compromise solutions are satisfactory provided that they work. In the case of safety, however, the effects of compromise can be very significant; it might be too late to introduce automatic gas detectors or prevent smoking in confined, unventilated spaces, for example, if a gas explosion has already occurred.

The provision of appropriate equipment and the setting up of suitable safe working procedures could be for nought if the principles are not implemented and practical advice and guidance not given on chemical safety matters. In addition, it is essential that equipment is properly maintained. All these factors readily allow the conclusion to be drawn that a designated, suitably qualified Environmental Safety Officer is a logical component of an environmental safety policy. It is then for the Safety Officer to make a judgement on the scale of his presence on a particular site, or part of a site, assuming that he has other sites to monitor. Where, for example, particular stretches of trenching or other excavations are through ground which, on the basis of historical and chemical investigation data, is considered to present a low hazard, then clearly he may consider that it is not appropriate to have a full-time presence. The site requiring the greatest care and protection is that where workers are least able to protect their own interests. Such sites are likely to have liquid and gaseous hazards and involve trenching and other excavation works.

8.6 Appendix: environmental health and safety guide for construction site staff (an example)

8.6.1 *Introduction*

All construction sites present some physical hazards and many workers are already familiar with safety precautions when, for example, working in trenches, using mechanical plant, handling inflammable fuel, etc. However, this particular site may present additional hazards with which site workers might not be familiar.

This site was previously used for a number of industrial processes which may have left residues in the ground such as solid or liquid chemical compounds which, together with other substances, might also cause gases to be formed. Some of these substances may be poisonous if inhaled or eaten, even in small quantities, and some may cause irritation to the skin. Such residues could exist anywhere on site but particularly in areas which will be specifically identified.

In order to help prevent accidents from such hazards, the site has been provided with a range of testing devices, protective equipment and emergency apparatus. There is also a full-time Environmental Safety Officer who will police all aspects of site safety as it relates to these hazards. However, personal safety is ultimately the responsibility of the individual.

Following the advice given in this guide will help you protect your interests more effectively. In particular, you must:

(i) Observe the Safe Working Procedures
(ii) Observe any instruction from the Resident Engineer
(iii) Be alert for the discovery of industrial residues in the ground.

In addition, normal construction site safety procedures must be followed, e.g. in the wearing of hard hats and boots with reinforced toe caps, instructions to machine drivers, and clearing debris or dangerous objects from working places.

8.6.2 *Safe working procedures*

8.6.2.1 *Safety Officer*. A Safety Officer (S.O.) has been appointed to help protect the safety of site personnel. The S.O. will:

(i) Regularly sample and analyse the working conditions
(ii) Advice on the appropriate handling or treatment of chemical substances encountered
(iii) Highlight particular hazards in particular locations at any given time
(iv) Give instruction on the protective clothing and safety apparatus to be used
(v) Organize and supervise a permit-to-work system.

It is essential that instructions from the S.O. are not treated as routine. They may vary from day to day, or hour to hour according to circumstances.

8.6.2.2 *Restricted Areas*. All new areas of trench excavation (and any other areas identified by the S.O.) will be classified as 'Restricted Areas' and only those workers with a 'Permit-to-Work' will be allowed to enter.

Permits will be issued each working day and must be returned at the end of that day. Permits will be restricted to a minimum number of essential staff.

No person will be allowed to work alone in a Restricted Area. In a trench designated as a Restricted Area at least one person must always be 'top-side'.

Smoking will not be allowed in Restricted Areas. Matches and cigarette lighters must be surrendered to the Safety Officer prior to the provision of a permit to work each day.

Work may not commence within any Restricted Area unless a Safety Equipment Box is close to hand.

8.6.2.3 *Personal safety and hygiene*. The following rules must be observed:

Thoroughly wash hands and face before handling food and eating.

Eating and drinking must be restricted to the rest room. No-one will be permitted to take any food (meal or snack) out on to any part of the site.

Trench workers must always wear overalls/boiler suit, hard-hat and wellington boots. (Other site workers may also be instructed by the S.O. to wear such clothing.) Workers may also be required to wear additional special protective clothing if instructed by the S.O.

If instructed, workers must wear safety goggles, face masks or special breathing apparatus when working within the Restricted Area.

Overalls worn by trench workers must be removed before leaving site and left in the changing room, unless otherwise permitted by the S.O.

Overalls should be laundered regularly, and changed immediately they become soiled with asbestos, tars, oils, solvents or similar material. Asbestos contamination will require specialist cleaning.

No protective clothing or equipment must be taken or worn off-site.

8.6.2.4 *Emergency procedures.* The following actions will minimize the consequences of an emergency:

In the event of any emergency or accident, no matter how small, sound the alarm immediately. The alarm shall consist of a 'continuous' intermittent high pitch signal. (This will summon the S.O. and warn others.)

If you start to feel drowsy or sick whilst working in a Restricted Area (or elsewhere) *get out*. Tell your mates to clear the area as well.

If a workmate collapses, summon help immediately and, if in a trench, get out, otherwise you may collapse yourself. (Wait on the excavation edge for help to arrive.)

Never go into a trench where someone has collapsed, unless you are (i) wearing a harness that is attached by rope to the digger and/or a line controlled by at least two workmates, and (ii) wearing suitable breathing apparatus.

Harness, breathing apparatus, masks, fire extinguisher and emergency first aid kits will be available in the Safety Equipment Box placed in the Restricted Area and at the main site office. Get familiar with this equipment and how to use it. The S.O. will provide instructions for its use.

8.6.3 *Form of chemical hazards*

Dangerous chemicals may be encountered in the form of: gases, liquids or solids (including dust).

8.6.3.1 *Gases.* Although most harmful gases are invisible, and some are odourless, they may have a rapid effect on you. Therefore, new trench excavations in particular should proceed with extreme caution. Signs to look out for are:

(i) Coloured fumes
(ii) Smell (e.g. bad eggs*, petroleum and other strong or unpleasant smells)
(iii) Drowsiness, dizziness or sickness

*The presence of a 'bad egg' odour should always be taken very seriously, since a high concentration dulls the senses, giving the impression it has disappeared. Therefore always advise the S.O. about *any* such odour as it presents a very serious risk.

(iv) Stinging eyes
(v) Tightness in the chest

Since some gases may be inflammable or explosive, smoking is not allowable within Restricted Areas.

If you suspect the presence of harmful gas, vacate the area and report it immediately to the S.O. who will carry out appropriate tests.

8.6.3.2 *Liquids*. Liquids from severed pipes and, perhaps, groundwater, are likely to form the principal hazards. Therefore, look out for accumulations or seepages of liquids which are:

(i) Tarry/oily in appearance
(ii) Fuming or bubbling
(iii) Brightly coloured (e.g. orange, blue, green) or darkly coloured (black/brown)
(iv) Strong-, oily- or sweet-smelling.

Avoid skin contact with such liquids and vapours. Report their presence to the S.O. who will check them and advise accordingly. (He may instruct you to wear special gloves and goggles while working with these liquids.)

Some of these liquids and their vapours may be inflammable. Therefore, do not smoke in the vicinity.

8.6.3.3 *Solids*. Solid contaminants are less likely to pose a problem, but if you find a new coloration in the soil that the S.O. has not seen that day, inform him.

The most significant material to be aware of is *asbestos*. Therefore, any greyish fibre/powdery material you discover must be avoided, and, if possible, sprayed with water. Secondly, inform the S.O. who will check the material and will instruct you accordingly.

You will threaten your health if you get contaminated dust or dirt on your hands and clothes and then eat food or go off-site without a thorough wash.

References

1. Health and Safety at Work Etc. Act, 1974, Chapter 37. HMSO, London.
2. Health and Safety Executive (1985) Guidance Note EH40: Occupational exposure limits. HMSO, London.
3. Health and Safety Executive (1984) Guidance Note EH10: Asbestos control limits measurement of airborne dust concentrations and the assessment of control measures. HMSO, London.
4. Department of Employment (1983) Statutory Instrument: S.I. 1983 No. 1649, Health and Safety: The Asbestos (Licensing) Regulations 1983. HMSO, London.
5. Health and Safety Executive (1984) Guidance Note EH42: Monitoring strategies for toxic substances. HMSO, London.
6. *Safetymate*. Labmate Ltd., St. Albans.
7. Health and Safety Executive (1981) Health and Safety (First Aid) Regulations SI 1981 No. 91. List of minimum contents for first aid boxes. HMSO, London.

9 Policy, planning and financial issues

R.C. HAINES

9.1 Introduction

It will be evident from preceding chapters that this book is largely concerned with the problems posed by contaminated land where such land is likely to experience a change of use or undergo redevelopment. The purpose of this chapter is to examine, from the perspective of the local authority, the measures that may be used to minimize any such problems. It addresses in particular the case for anticipatory action within the framework of local planning procedures and the actions that may be taken either to avoid the unsafe development of contaminated land or bring about a satisfactory site investigation. Thus, it outlines the planning procedures open to local authorities in dealing with potentially contaminated land. It goes on to discuss the role of building regulations and the powers available under public health regulations and development control to bring about the safe and economic redevelopment of such problem sites.

The chapter goes on to review the role which public sector financial support in the UK, in the form of derelict land grants or other assistance, might play in treating contaminated land and bringing it forward for development.

9.2 Controlling the development of contaminated land

It has already been pointed out in Chapter 1 that the UK approach to contaminated land operates almost exclusively in the context of contaminated land redevelopment.

The increased pressure for the re-use of land, coupled with a long history of industrial development in the UK, has encouraged the re-use of former industrial land for other purposes. The redevelopment, and hence revitalization, of former industrial and inner-city areas, where much of the contaminated land is located, is generally considered to bring a number of social benefits.

UK policy in respect of contaminated land is therefore to promote the redevelopment of contaminated land for beneficial purposes. Local authorities and others with responsibility for contaminated land are encouraged to adopt a systematic approach to identifying and assessing any hazards associated with contaminated land in relation to the redevelopment use. Information derived at each stage of investigation and assessment is then used to decide whether a redevelopment proposal goes forward unmodified, whether remedial measures are required or whether an alternative use of the site would be more appropriate.

The discovery that a site is contaminated during its development can be costly, particularly where it may not be possible to change the nature of the development and/or there are long delays because of the need for site investigation work and ameliorative treatment. There is therefore a need to ensure that contaminated land is taken into account at an early stage in the development process.

The statutory and legal powers available to local authorities when exercising their responsibilities for planning and development control potentially contaminated land; they include ensuring that sites are developed in a way which minimizes the risk to public health or to building structures.

There is no specific guidance on controlling the development of contaminated land in the UK, although some control can be exercised through existing planning, public health and safety legislation (1–7).

9.3 Land-use planning and control

There are three principal areas where local authorities could potentially respond to the problems of contaminated land. These are

(i) Forward planning
(ii) Development control
(iii) Application of building regulations.

9.3.1 *Forward planning*

Forward planning is the procedure by which local authorities control the future use of land; it allows developers to find out where, and what type of, development is likely to be permitted in an area. It is undertaken largely in the framework of Structure Plans and Local Plans.

Structure Plans are prepared by County Planning Authorities, and approved by the Secretary of State after an examination in public. They concern certain matters of 'strategic' importance, state broad land-use policies and general proposals for the development and other uses of land. They are not intended to be site-specific or to 'zone' land. They identify needs: e.g. the number of houses required, and they provide the framework within which detailed local plans are formulated, and development control is administered. However, the Structure Plans of different County Councils provide different amounts of information. For example, the structure plan for the West

Midlands Metropolitan County includes details on large sites which are allocated to specific uses, e.g. housing or industrial uses. Others do not. It is considered that structure plans may be of limited utility in terms of ensuring contaminated land is identified and investigated prior to redevelopment.

Local Plans may be prepared by County or District authorities, but most are prepared by the districts. They have four main functions:

(i) To develop the policies and general proposals of the Structure Plan, and relate them to precise areas of land
(ii) To provide a detailed basis for development control
(iii) To provide a detailed basis for co-ordinating the development and other uses of land
(iv) To bring local planning issues before the public.

They thus determine detailed land-use policies for specific areas and sites, and provide developers and others with clear guidelines of where development is most likely to take place. They may define a specific site for development or other use, or define the specific areas within which particular policies will be applied: e.g. areas selected for comprehensive treatment by development, redevelopment or improvement.

A Local Plan should comprise

(i) A written statement, setting out and justifying the policies and proposals for the development and other uses of land
(ii) A proposals map, defining sites for development and other land uses, and the sites where particular policies will be applied
(iii) A land use diagram, showing a long-term picture of the principal existing and proposed land uses.

The Town & Country Planning (Structure and Local Plans) Regulations, 1982, provide specific guidance on the format of 'local' plans. Typically the area covered by a plan is the whole of a small town or a large part of a big one. Where the Local Plan is concerned with only one subject, for example minerals, (and called a Subject Local Plan), it will usually be prepared by the County for the whole area. Action Area Plans are local plans which cover relatively small areas, e.g. from one or two streets up to a ward in size, which are subject to intense change, such as a housing review.

The largest land-use changes are most likely in areas which have fallen into decay (inner-city areas and major industrial closures, such as steelworks, docklands, etc.), and which now offer opportunities for improvement by the creation of residential, commercial or recreational uses. It is probably in these areas that there is the greatest amount of potentially contaminated land. Such local plans should, according to DoE Circular 4/79, take into account physical as well as economic and social aspects of the area. In this context, awareness of potentially contaminated sites or problem sites could assist in judging whether land is suitable for a particular development, and whether or not proposals could be realized in the time-period. This would avoid proposals for sites which could not be realized in the time-scale because they would require investigation and possible ameliorative treatment before development could proceed. It could also 'flag' potential problem areas, particularly areas identified for housing schemes (industrial or commercial uses are likely to be less sensitive) where further, detailed investigation might be required.

The Local Planning Authority (LPA) could respond to the issue of potential contamination in three ways in the context of such a local plan.

(i) It could ignore reference to the possible presence of contamination by allocating land uses irrespective of the condition of the land. This is considered unrealistic because of the responsibilities and liabilities of LPAs under development control and building regulations as discussed below.

(ii) It could identify areas where potential land contamination problems are thought to exist and to make no allocations in those areas until detailed investigations have taken place. This approach would, however, seriously affect the prospects for the regeneration of decaying inner-city and industrial areas. Existing properties could become unsaleable and the area could be blighted. It is also conceivable that owners of sites in such areas might, if refused planning permission for development, conclude prematurely that sites have no reasonably beneficial use and serve purchase notices on the LPA.

(iii) It could allocate land uses in areas where contamination is potentially a problem, but make it clear in the plan that development must await detailed site investigation and if necessary ameliorative treatment, thus forewarning prospective purchasers and developers of problem sites well in advance of either a commitment to development or the drawing up of a detailed planning application. This requires that LPAs have detailed knowledge, not only of present and past uses or sites but also, and more importantly, evidence on which to base judgements. The reservation in adopting this approach is that the land-use allocations may not meet local plan needs, requiring that the total land allocated to some of the more sensitive uses (e.g. housing and schools) is greater than for other uses such as industry and open space. Such an approach could well sterilize land capable of beneficial use. The general consequence might be that the local plan would no longer be a workable planning document.

9.3.2 *Development control*

Development control and the determining of planning applications occur, in an 'ideal world', possibly too late in the land-use planning system to take proper account of potentially contaminated sites. Nevertheless, guidance is required on how best to deal with planning applications of contaminated sites. With certain exceptions, all development requires the prior approval of the LPA. Development is defined in section 22 of the 1971 Town & Country

Planning Act (TCPA) as including both 'building, engineering, mining or other operations,' and the 'making of material change in the use of any buildings or other land'. The LPA, when determining a planning application, is required to have regard to the provisions of the development plan and to any other material considerations. It can, however, approve a proposal 'which does not accord with the provisions of the plan', if the proposal does not involve substantial departure from the plan and does not injuriously affect the amenity of the adjoining land.

When considering a planning application the LPA should consult and co-ordinate with other relevant authorities and government departments, and have regard to the development plan (Structure Plan or adopted Local Plan). Section 29 (i) of the 1971 TCPA also requires the LPA to have regard to any 'other material considerations'. On receipt of an otherwise acceptable application, the LPA can take one of two courses of action, where contaminated land is involved: (i) regard the issue of site 'contamination' as not being a planning consideration and determine the application accordingly, or (ii) regard the issue of contamination as a 'material consideration'. Three decisions are possible: refusal, permission subject to conditions, and unconditional permission. Any decision must be critically dependent upon the quality of the information available.

(i) *Refusal.* Planning authorities may regard contamination as a material consideration; that is they may refuse planning permission if they know that the site is contaminated and therefore unfit for its intended purpose. Nonetheless, the LPA is not responsible for ensuring that the physical condition of the site is suitable for the proposed development in determining an application; can this, therefore be true of chemical suitability?

However, the refusal of planning permission on the grounds that the site is contaminated is liable to reduce its value and that of other comparable land in the area, and could well make it impossible to sell. In consequence, the owner could serve a purchase notice on the LPA under section 180 of the TCPA 1971 on the grounds that the site had become incapable of reasonably beneficial use in its present state. If a purchase notice were accepted or confirmed by the Secretary of State, the LPA (or such other statutory authority or statutory undertaking as might be substituted for it) would be obliged to acquire the land. The purchase price would not necessarily be negligible, because the site would have to be valued in accordance with statutory assumptions, and not at its market value.

(ii) *Planning permission subject to conditions.* Planning permission, subject to conditions, can be a very useful mechanism where there are uncertainties. The legislation allows the LPA to grant permission subject to such 'conditions as they think fit'. However, where conditions are attached they must be fair and reasonable. Although general guidance concerning 'conditions' was given by a

Circular (12) it contains no specific advice on contamination. A new Circular on 'conditions' is in preparation and the Department is considering whether this would be an appropriate vehicle for advice on the use of conditions in relation to problems of contamination.

Although the LPA is not responsible for ensuring that the physical condition of the site is suitable for the proposed development, where they are aware, or have reason to believe, that the site is contaminated and they consider this contamination to be a material consideration (though not one warranting refusal of planning permission), it would seem that two courses are open to them. The first is that permission could be granted, subject to a condition that the applicant carries out an investigation of the site before development begins, in order to satisfy himself that the land is suitable for the development proposed. The other is that the permission could be accompanied by a letter from the LPA explaining why they think it is in his own interest that the applicant should carry out such an investigation.

(iii) *Unconditional permission.* If the LPA decides not to regard contamination as a material consideration, the granting of planning permission without a condition in that respect or without an accompanying advisory letter may be misleading to the applicant about the true prospects for development.

It is undesirable for any LPA base planning permission on 'suspicion' obtained from a desk study that contaminants may be present on a site, leaving them uncertain as to the extent of the contamination and whether there is any substantial risk. This reservation applies to almost all decisions in the absence of guidance. It applies also to site investigations and to their interpretation, including how far LPA officials have sufficient expertise (or guidance) to assess that risk of contamination, and to balance those risks against the cost of a site investigation and treatment against actions leading to unnecessary blight.

9.3.3 *Role of building regulations*

Building Regulations[1] are made under the Health & Safety at Work Act 1974 in England and Wales (separate provisions exist in Scotland and Northern Ireland). They are a possible mechanism for controlling the development of contaminated land, but offer relatively limited powers. Most buildings and structures (with few exceptions, e.g. car and lorry parks and sports grounds) require approval under the Building Regulations to ensure the health and

[1]The Building Regulations for England and Wales have, in the past, not applied to the part of London covered by the London Building (Constructional) By-laws 1972, and the powers open to the District Surveyors in Inner London have been different from those available to Building Control Officers elsewhere. However, after the new national Building Regulations came into effect on 1st November 1985, the London Building (Constructional) By-laws 1972 were repealed and replaced with the new Regulations.

safety of persons in or about buildings. The principal regulation is D3 (C) which states that 'foundations shall be capable of resisting any attack by sulphates or other deleterious matter in the sub-soil'. District Councils administer building regulations, and where they are not satisfied about the suitability of the ground conditions, they may refuse the application. The applicant has then to show that the site is not affected or that his proposals will overcome the problems by ameliorative treatment.

Local authority current practice appears to be reasonably consistent. The building surveyor will require that soil samples on suspect sites are tested for sulphate and bearing capacities. The decision is normally based upon local knowledge and experience of dealing with problem sites. The powers are limited; although the regulations cover sulphate attack on concrete and possibly organic attack of service pipes, they do not extend to environmental health problems. There are no Building Regulations at present for metals. Many local authorities are of the opinion that potential problems should have been identified and dealt with at a much earlier stage.

It would be unfortunate were these applications to be refused under the Building Regulations. If, in extreme cases, a Council were judged to have failed to have taken responsible care, i.e. to have ignored possible ground condition problems, and a building or structure later failed due to adverse ground conditions, there are legal precedents which can mean that the authority would be liable to a civil claim for damages for breach of statutory duty and negligence. At present this is limited to failures of building structures due to attack by contamination, but the Council might also be liable for compensation to individuals who suffer ill-health as a result. Thus, attempts to control the development of contaminated sites using the present Building Regulations may well come too late and generally be too complicated and uncertain.

There is a new requirement in the Building Regulations (9) that building sites should be assessed for contamination and appropriate remedial action taken. The main difficulty with this provision is that it only covers the land immediately beneath a building, and does not recognize the possibility of migration of contaminants, especially gases, and as discussed in Chapter 11.

9.4 Public health legislation and development control

Powers are available under the Public Health Act (1936), mainly sections 54 and 92, allowing local authority environmental health departments to control the development of contaminated sites at the time of a planning application for proposed new development, so that they may ensure that public health is not put at risk due to the presence of contaminants.

Section 54 of the Act deals with ground filled with offensive matter, and empowers local authorities to reject applications under the Building Regulations if faecal or offensive animal or vegetable matter is present, unless it has been rendered innocuous. It has therefore only been applied to landfill sites, especially those where methane generation (domestic refuse) may be a problem or infected animal matter has been deposited. It is evident that many authorities find section 54 too vague, and are therefore unwilling to apply it to contaminated sites other than landfill sites. This is particularly true for other hazardous materials, such as toxic chemicals or heavy metals, since only after a detailed chemical site investigation can the nature, intensity and distribution of contamination be determined. The section also omits to specify who should do this work and who must pay.

Section 92 of the 1936 Act, which should also be read in conjunction with the Public Health (Recurring Nuisances) Act 1969, supplements the common law of public and private nuisances by creating the concept of a 'statutory nuisance'. The definition relates to (i) any premises in such a state as to be prejudicial to health or a nuisance; (ii) any accumulation or deposit which is prejudicial to health or a nuisance; or (iii) any dust or effluvia generated by trade, business, manufacture or process being prejudicial to health, or a nuisance to, the inhabitants of the neighbourhood. 'Premises' include buildings, land or easements. If a local authority is satisfied of the existence of a 'statutory nuisance', then it can serve an abatement notice, provided that there is proof of injury to health or a nuisance. The notice can specify any necessary action to prevent the nuisance. Under the 'recurrence of nuisance' provisions in the 1969 Act, when a local authority is satisfied that a statutory nuisance has occurred on any premises or is likely to recur on the same premises, they may serve a 'prohibition notice' which can specify the work necessary to remove it. Although contamination may fall within the definition of a statutory nuisance, there are a number of difficulties in applying these provisions. No nuisance can exist until the operations at the premises give rise to them. They cannot therefore be used to prevent the unwise development of a contaminated site (e.g. new housing development on a gasworks site). They can only be used to ameliorate a problems once the development has been completed. Moreover, it may be many years before the health of occupants of buildings on contaminated sites is adversely affected, and a detailed site investigation would be required to establish the exact cause of any such effect which may have occurred.

9.5.1 *Derelict land grant scheme*

Derelict Land Grant provides finances to reclaim sites which otherwise would remain derelict or a nuisance. Section 89 of the National Parks & Access to the Countryside Act (1949) as substituted by section 3 of the Derelict Land Act (1982) confers specific powers on county and district councils to carry out works to enable derelict, neglected or unsightly land to be reclaimed, improved or brought into use. Provision is also made in the 1949 Act for local authorities to acquire land for such purposes, by agreement or compulsorily. The various other powers under which local authorities can acquire and develop land for particular purposes can, of course, be exercised in relation to derelict land.

Successive governments have been aware of the problems created by derelict land, and in 1966 legislation was enacted enabling grant to be paid to local authorities for reclamation works. Although no statutory definition exists, for grant purposes the DoE has agreed with the Treasury that derelict land is defined administratively as 'land so damaged by industrial or other development that it is incapable of beneficial use without treatment'. Thus, vacant industrial land which is potentially contaminated can be put forward for derelict land reclamation.

In England, central government grants are available under section 1 of the Derelict Land Act 1982. The level of grant available for reclamation by local authorities of a given site will be subject to the location of the site in relation to a designated Assisted or Derelict Land Clearance Areas, including English Industrial Estates Corporation, where 100% grants are available. Elsewhere, the percentage of works grant available will be 50%, except in the case of National Parks and Areas of Outstanding Natural Beauty where 75% grant is available. As far as local government block allocations are concerned, derelict land grant does not count against Other Services Block. Hence apart from the limits imposed by the Regional Association there is no financial constraint to local authorities' undertaking land reclamation.

Over the past few years a number of significant changes have taken place in the derelict land grant system. The Local Government Planning and Land Act 1980 extended the provisions for grant to be made available to 'any person', such as private companies, nationalized industries and statutory undertakers. The grant is 80% in derelict land clearance areas and 50% elsewhere. Then, in late 1981, a new initiative was announced by the DoE which introduced a two-tier system to give priority to local authority schemes which would be followed by private sector development. These are known as category A schemes, and within this category, preference is given firstly to industry, then to commercial, recreational and other forms of development.

The second tier, known as Category B schemes, are those promoted and developed in their entirety by local authorities. In addition, greater emphasis is now placed on urban areas, with priority being given to applications from Partnership Authorities and Other Authorities designated under the Inner Urban Areas Act 1978.

The important feature of this 'new initiative' is that derelict sites with the right potential for development could be reclaimed and developed under a joint venture between local authorities and the private sector. This can be achieved through one of four ways:

(i) The local authority acquires the site and restores it to a 'green field' condition ready for development, the developers buying the site at market value

(ii) The developers contract in advance of reclamation to buy the land from the local authority after reclamation has been completed

(iii) The developers carry out the reclamation under contract from the local authority who owns the land and who passes it over to the developer once reclamation has been completed

(iv) The developers buy the land and carry out the reclamation work themselves.

Costs incurred by the local authority in (i)–(iii) are supported by derelict land grant at 100% of eligible expenditure in Assisted and Derelict Land Clearance Areas. Costs incurred by the developer under option (iv) would be eligible for 80% grant aid. Derelict land grant is payable on any net loss incurred in carrying out the reclamation work.

If, therefore, a reclaimed site is sold, leased or appropriated for development purposes (other than amenity or open space which is considered to have nil value), then the value of the land after reclamation, or the increase, is set off against the grant entitlement. In instances where this applies, grant is paid in full for reclamation and after-value is paid by the local authority at a later date when the land is actually transferred and the value accrues. Before 1982, after-value was always required to be paid in full to the DoE upon realization of value by the local authority. The two-tier measures for preferential treatment instigated in 1981 did, however, provide for special arrangement for category A schemes, with local authorities being allowed to retain up to 50% of the after-value of a site in all but the most profitable ventures. Where this occurs, the retained sum can be spent on other reclamation sechemes (either Category A or B).

9.5.2 *Works eligible for finance*

The underlying philosophy behind the grant-aiding of reclamation is that once reclamation has taken place, further development should be no more difficult than on a greenfield site. This concept has had an important influence on the scope of the works which are eligible for grant aid, particularly in the context of contaminated land. The most important eligible works are the provisions for the demolition and removal of unwanted buildings and plant and site clearance works which includes for the treatment and/or removal of industrial and other wastes including contaminated material and any necessary works to treat hazards (e.g. extinguish spontaneous combustion, deal with methane generation, etc.).

More importantly, the approved costs in carrying out a survey of derelict land are eligible for grant. Survey in this instance includes topographical surveys, ground investigations, structural surveys to assist in the preparation of demolition contracts and environmental surveys. The grant is payable whether or not any reclamation works are subsequently carried out. Thus, for example, a former industrial site being considered for housing development where contamination is suspected, in an inner-city area, may be eligible for grant aid to carry out a detailed physical and chemical site investigation to prove the condition of the ground. Where such sites are in assisted or derelict land clearance areas, the full costs of the site investigation can be met out of grant aid. In addition, for surveys and works, local authorities may claim 10% enhancement of the overall cost for administrative expenses incurred during the design and supervision of reclamation schemes.

Other eligible works include earth moving, land drainage essential to the works of reclamation, accommodation works, and ground preparation and topsoiling.

With regard to infrastructure, for category B schemes, grant aid may be payable on the costs of providing both on- and off-site infrastructure where the site is being reclaimed for industrial, commercial or residential use. In the case of category A schemes, 'on-site' works such as site compaction, foundations and reservicing would have to be financed by the developer concerned.

From the above, it will be evident that there exists a reasonably comprehensive system for ensuring that contaminated derelict land is brought back into beneficial use in a safe and cost-effective manner. Add to this, the provision of grant aid for site investigation works, and there appears to be a mechanism through which sites can be assessed prior to sensitive developments taking place and any contamination hazards identified dealt with.

The biggest stumbling block is the level of resources available. The Government has increased the resources available for reclamation in England from about £23 m in 1979/80 to something over £76 m in 1985/86. The private-sector schemes take under 10% of these resources. This level of resources is considered by many local authorities to be wholly inadequate in relation to the scale of the derelict land problem generally and, in certain parts of the country, to contaminated land in particular (e.g. West Midlands and North-West). Other problems relate to the operation of the system. The scheme appears to be unattractive to the private sector because of the lag between initial approach and grant offer. Indeed, grant offer may be subject to an applicant waiting in a queue until resources are available. With regard to the public sector schemes, the greatest difficulty is reconciling the need for a rolling programme to implement priorities with an annual cash limit and a DoE desire to keep year-on-year commitments to an absolute minimum.

Other public sector resources. A significant and growing proportion of Urban Programme (UP) expenditure, which provides resources for local authorities for environmental improvement and economic development in inner-city areas, has been directed at environmental improvement projects. Central

Government provides 75% grant support for approved expenditure with local authorities finding the remaining 25% from their own resources. In 1983/84, a total of some £47 m of UP expenditure was allocated to environmental improvement projects out of some £315 m. Not all local authorities are eligible for Urban Programme funding. The original list of beneficiaries of Urban Programme funding has been amended in the light of changing circumstances, and a widening perspective of the inner area problems. The Local Government, Planning and Land Act 1980 redefined certain areas, whilst the Secretary of State for the Environment, in announcing New Urban Initiatives in February 1985, amended existing and introduced new designated areas—Partnerships, Programme and Other Designated Districts. The Partnerships were set up to co-ordinate all the agencies involved in urban renewal. The local authorities prepare a three-year rolling programme of schemes for Urban Programme funding.

A wide range of environmental improvement projects can be supported out of UP resources. These include landscaping, the provision of parks and open spaces, general environmental works and special improvement projects. In 1984/85, the Partnerships spent some £22 m on reclaiming derelict sites, which produced some 220 sites with firm end-uses, 1025 acres for industry and 85 acres for housing. Many of these sites were former industrial land, some of which were contaminated

9.5.3 *Urban development grant*

Urban Development Grant (UDG) aims to promote the economic and physical regeneration of inner urban areas by encouraging private sector investment. UDG is available for capital regeneration of inner urban areas by encouraging private-sector investment. UDG is available for capital investment projects which would not go ahead without some public-sector assistance, but which, with that assistance, are commercially viable and help to deal with urban deprivation. Not all local authorities are eligible for UDG support. It is normally available in areas where there are severe problems of urban deprivation, such as unemployment, bad housing or derelict land and buildings.

Any size or type of scheme can be put forward for UDG. In general, projects funded under UDG are similar to category A, DLG schemes. However, there is one important difference, in that all UDG require substantial private-sector capital investment and must provide social benefits such as new jobs, good-quality housing or reclamation of derelict sites or buildings. There is no formal guidance on the level of support available from the public sector. The private sector has to meet the greater part of project costs. But, if the value of the completed development would be insufficient to provide an adequate return on total project costs, UDG is available to bridge the gap. The amount is decided after a detailed appraisal of the project. UDG may be a grant or a loan or a combination of both. If the circumstances justify it, loan interest may be waived or grant may be conditionally repayable. Most UDG schemes are undertaken by companies and schemes approved include shop, factory and office building, house building and refurbishment and the provision of leisure and hotel facilities. To date, some £65 m of grant has been approved for expenditure on projects involving over £290 m of private sector resources. UDG can and indeed has been used to 'bridge the gap' to allow the successful redevelopment of contaminated land.

A good example of such schemes is Holford Park in Birmingham, a former IMI site, heavily contaminated from a beryllium works, waste disposal site and an area formerly occupied by operations relating to explosive manufacture. IMI put forward redevelopment proposals which were to create a high-quality engineering park of some 18.2 ha. IMI received a UDG of £4.9m, which is the largest grant provided to date, some of which was required to deal with the contamination problems. For example, the demolition/decontamination of the beryllium works cost about £90,000 and took 4–5 months, compared to an ordinary demolition time of 4 weeks. On the waste disposal site, where extensive areas of asbestos and copper sludges were present, remedial treatment cost about £260,000.

References

1. Town and Country Planning Act, 1971.
2. Health and Safety at Work Act, 1974.
3. Public Health Act, 1936.
4. Public Health (Recurring Nuisances) Act, 1969.
5. National Parks and Access to the Countryside Act, 1949.
6. Derelict Land Act, 1982.
7. Local Government Planning and Land Act, 1980.
8. Inner Urban Areas Act, 1978.
9. London Building (Construction) By-Laws, 1972.
10. Building Regulations, 1985.
11. Department of the Environment Circular 28/85.
12. Department of the Environment Circular 4/79.
13. Ministry of Housing and Local Government Circular 5/68.
14. JURUE: Identification and Assessment of Contaminated Land. A report for the DoE, ECOTEC Research and Consulting Ltd., Birmingham, 1984.
15. Cullingworth, J.B. (1976) *Town and Country Planning in Britain.* George Allen and Unwin, London.

10 Landscaping and vegetating reclaimed sites

G.D.R. PARRY and R.M. BELL

10.1 Introduction

Other chapters have described the complex nature of land contamination and as a consequence of this, the variety of remedial actions available for reclamation of contaminated land sites. Similarly, when considering landscaping and vegetating as the final process in reclaiming sites, no single definitive approach can be identified. The landscape specification for each individual site and the response of vegetation to the conditions prevailing will of course depend primarily on the way in which the site was reclaimed in the first place.

In the context of assessing landscaping requirements and limitations, it is possible to identify two main categories of contaminated land treatment, which are:

(a) Removal of the waste or contaminants and backfilling with clean material.
(b) Treatment in which the contaminated material is left in place and either treated to render it inert or containerized.

The latter, which allows the contaminant to remain on site either in an altered inert form or isolated by covering or encapsulating layers, is the most common. Landscaping and vegetating sites treated by these two broad methods will have some aspects in common, but others will relate particularly to the second treatment type.

On sites which have been cleared completely the normal problems of establishing vegetation on juvenile or poor quality soils, or on soils which have suffered from excessive handling, will apply. These aspects will also apply to sites in which the contaminants remain, but in addition the important influences of soil/plant relationships must be considered in relation to maintaining the integrity of the treatment system. Thus where the contaminant remains on site, in one form or another, questions of contaminant uptake by plants and the impact of plant roots on the integrity of covering and encapsulating systems must be reviewed, in addition to the requirements of the plants themselves.

10.2 Requirements of plants

Development of a successful landscaping scheme will depend in part upon sympathetic treatment by the landscape designer, but will not succeed unless the basic needs of plants to sustain growth are clearly understood. These needs, as stated above, are common to all reclamation schemes. Differences arise when plant behaviour may in fact influence the reclamation, or may be used as an integral part of the reclamation scheme.

In the absence of conditions toxic to plants, there are four main conditions that need to be satisfied to achieve good vegetation growth whether in grasses, shrubs or trees. A plant requires sunlight, the fuel for photosynthesis, physical support, water, and plant nutrients. Their provision must form part of the design and reclamation specification.

10.3 Physical support for plants

Nearly all plants used in land reclamation scheme landscaping require support by anchoring their roots in the growing medium. This is achieved by the roots

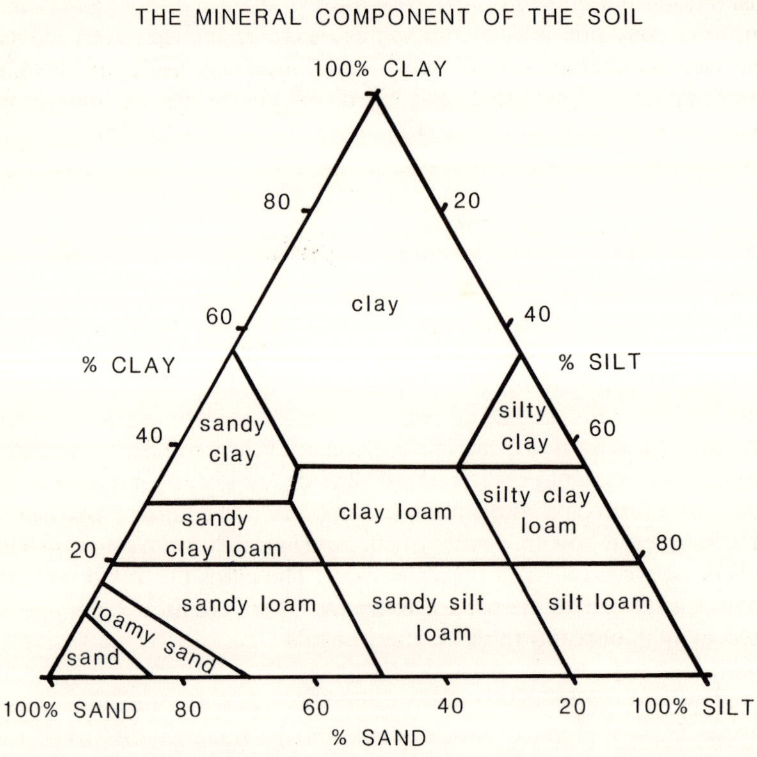

Figure 10.1 Soil textural triangle: this demonstrates the classification of soil types according to their relative percentages of sand, clay and silt.

growing around small particles in the substrate. In physical terms alone the degree to which plants can accomplish this is dependent upon the substrate texture. Soils and substrates can be classified according to their relative percentages of sand, silt and clay-size particles. When the percentages of each fraction have been determined then the soil or substrate can be classified in terms of texture according to the texture triangle (Figure 10.1), which has been developed for agricultural purposes (1).

Physically there are a variety of materials available which, with amendments, are suitable for growing plants. The most commonly used material is, of course, topsoil, largely because of its additional merits of containing plant nutrients and providing adequate water holding capacity to keep rainfall in the vicinity of plant roots. This material can therefore aid the fulfilment of three of the four conditions required by a plant for good growth, the exception being the provision of sunlight.

Topsoil is expensive and takes thousands of years to develop naturally. Careful handling and storage are required if this material is to be used for reclamation and landscaping, in order to preserve its beneficial qualities. Once the soil structure is destroyed by poor handling, then many of the advantages of using topsoil are also lost (2).

In many cases there is little or no topsoil available as a final cover on treated sites. In such instances selected subsoils are often very useful in providing a support medium. These are the soil layers that lie beneath the topsoils, often down to the bedrock within a soil profile. In general they have a poorer physical structure than topsoil and contain less plant nutrients. Supplies are usually obtained from local civil engineering projects which involve earthworks, but they must be handled and chosen with care. Heavy clay subsoils, for example, are extremely difficult to handle and generally provide a poor plant substrate.

Where topsoil or subsoil are unavailable it may be possible to substitute a waste material of suitable physical properties in their place (3). Obviously any such material would require examination to ensure that it itself did not possess phytotoxic properties. Widely used examples include waste from sand winning, pulverized fuel ash (fly ash), colliery spoil, composted refuse and sewage sludge. Apart from the latter example, all the others quoted would require some form of amelioration to ensure that the requirements other than root support for plant establishment and growth were met, including provision of plant nutrients discussed later. The latter two materials should always be analyzed before use to ensure that there are no toxicity problems caused by high concentration of heavy metals.

10.4 Compaction

For a plant to obtain adequate physical support for growth depends not only on soil texture, as discussed above, but also on soil structure. This is determined by the way in which soil particles come together to create larger particles separated by pore spaces. When soil or other substrates become highly compacted as often happens when heavy machinery is used for ground preparation, the pore spaces are reduced and bulk density of the substrate increases.

Reduction in pore space seriously restricts the ability of a substrate to drain. Some substrates (e.g. the sandier varieties) retain their structure better than others. Wet clay rich materials, however, tend to become very compacted with little pore space. When this occurs and drainage is impeded, then the availability of air to the roots is reduced, inhibiting root growth and the development of the substrate into a biologically active soil.

10.5 Water supply and storage

All plants require water to grow and this is normally supplied by rainfall or early morning dew. Thus as the supply is not continuous plants need a storage system where water can be held for latter use when external supply is short. In temperate zones the water storage system is supplied by water held between the clay sized particles of the soils and it is their presence which largely governs the quantity of water that can be stored and the ability of vegetation to withstand drought. It follows, therefore, that sandy soils can store little water while clay soils can store considerable quantities of water. A good quality topsoil will have good water storage capacity but this is not necessarily true for subsoils or waste materials used as growing media. These may be deficient in clay content or contain excess clay fractions.

10.6 Plant nutrients

Plants need a continuous supply of nutrients from the growing medium both during the establishment phases of a landscape scheme and for continued growth. The major plant nutrients are nitrogen, phosphorus and potassium, together with a number of positively charged ions, including calcium, magnesium and sulphur. A further group of elements are the trace elements which are only required in very small quantities by plants.

When topsoil is used in landscaping schemes a proportion of these nutrients will be supplied by the natural nutrient content of the soil itself. When sub soils or waste materials are used serious deficiencies of the major plant nutrients may occur which will have to be added in the form of fertilizer amendment to ensure plant growth.

Under established vegetation cover nutrients continue to recycle within the growing medium and the plants themselves. Simply, plants take up nutrients, use them to grow, die, break down within the soil and release the nutrients ready to be taken up again. When topsoil is moved from one site to another, the established vegetation is removed and lost and because the nutrients within the vegetation are removed the cycle is broken. Similarly, when topsoil is stored for long periods in saturated or compacted conditions, as might occur in high, large topsoil heaps, severe and irreversible physical changes may take place. These generally lead to the onset of anaerobic conditions at the centre of the stockpile and also result in a loss of fertility.

A large input of fertilizer is required to establish vegetation but little input is required once it is established and nutrient cycling is occurring (4).

When new grass swards are being established onto topsoil from seed in the spring a typical application of nutrients should be made by the addition of up to 300 kg/ha of a proprietry brand of 17:17:17 compound inorganic fertilizer. This provides the equivalent of 51 kg/ha of nitrogen, 23 kg/ha of phosphorus, 42 kg/ha of potassium. A further similar dressing should be applied during the autumn of the same year. Applications of lower dressings of fertilizers equivalent to two thirds the initial dressing should be applied annually for at least the first three years growth to ensure establishment of the nutrient.

When subsoils or waste materials are used for vegetation establishment the number of dressings of fertilizer should be increased to two heavy dressings of fertilizer for the first three years of grass establishment. If new grass is to be established in the autumn then the initial dressing of fertilizer should contain lower amounts of nitrogen in order to prevent excessive new growth which could be damaged by winter conditions.

Some organic fertilizers are available that also contain nitrogen, potassium and phosphorus and which can be used in place of inorganic fertilizers. They include sewage sludge and chicken manure, and as they are often waste products, they are very cheap. Their use is not as simple as inorganic fertilizers and expert advice is often required.

Some of the nitrogen requirement of a grass sward can be provided by the inclusion of legumes, such as clover, within the seeds mix. Legumes have the ability to convert atmospheric nitrogen into nitrate for use for growth. Most importantly, however, legumes do not use all the nitrate they produce for themselves and some enters the soil for use by adjacent plants.

Trees and shrubs also require fertilizers. The larger the tree or shrub being established, the more fertilizer it requires. Once established, however, nutrient cycling occurs and no more provision of fertilizer is required. When planting in pits or notches, fertilizer should be applied at a rate of 70 g of 10:15:10 compound fertilizer per tree to be incorporated into one cubic metre of soil in the rooting zone. This application should be repeated each year throughout a 3-year establishment period but spread 1 m^2 around the plant.

Some trees and shurbs also have the ability to fix atmospheric nitrogen for use for growth. These plants include alder, gorse and broom and are therefore very useful when reclaiming or revegetating land. Their use as a nurse crop to protect slower growing trees and shrubs on nutrient poor substrates can be extremely successful.

10.7 Grassland establishment

Production of grass swards invariably constitutes a major part of the revegetation or landscaping proposals for reclaimed sites. The methods used to establish grassland are dictated by the condition of the site and its intended use. When earthworks have been carried out to change topography or provide amendment cover to a contaminated site, the action of machinery generally results in soil or cover compaction. This must be alleviated in order to form a seed bed. In highly compacted soils, ripping to at least 150 mm using 300 mm tines on 600 mm centres is required. On less compacted soils, mouldboard or chisel ploughing is often satisfactory. It is obviously not advisable to cultivate at depths which might disrupt a covering system.

Following the alleviation of compaction, stone-picking is often necessary, carried out by hand or by specialized machinery, depending on the size and topography of the site. The size of the objects to be removed also depends on the final use of the site. On land restored to close mown grassland, all small stones (> 50 mm) must be removed but for rough grassland, only large stone (> 100 mm) must be removed.

The initial cultivation and stone-picking often leave a rough surface which is unsuitable as a seedbed. This can be worked into a finer tilth using a disc harrow after fertilizing in order that these materials can be incorporated into the surface layer. Following this cultivation, a single pass with a Cambridge roller then helps to consolidate the seedbed. These methods of cultivation are only applicable on many soils when the weather conditions are suitable. For example, if the land lies wet for long periods, it is virtually impossible to produce a good seedbed of fine textured soils. The timing of operations is therefore crucial to the success of the scheme.

10.8 Grass seed mixtures

The seeds mixture used must be chosen on the basis of soil conditions, the location of the site and the final use. Grassland for public open space or sports uses must be durable and attractive. Such qualities are not consistent with those of agricultural pastures, which must be highly productive, with grasses of high nutritive value. Grassland for agricultural use may be established using seeds mixtures and procedures commonly used for producing long term leys.

In some instances it may be necessary to consider the use of hardy grass cultivars and varieties for revegetation of less hospitable sites. These can withstand more extreme environmental stresses. Such cultivars may include those which have been developed because of their genetic tolerance to stress, for example, some heavy metals or drought. The availability of such cultivars should always be considered. If available, they would normally replace the cultivar of the same species included in the seeds mix.

It is often difficult for plants to establish and grow on reclaimed sites and grass seeding rates should be 30–50% more than those for 'normal' soil conditions.

10.9 Grass cutting

The frequency of mowing depends on the type of sward and the final use of the reclaimed land. For intensively used sports and recreation areas, the sward should generally be mown once every two or three weeks during summer. Many small cuts are better than one big cut. Attention must be paid to the condition of the vegetation, and the frequency of mowing should be adjusted accordingly. In periods of drought or waterlogging, frequent mowing will damage the sward.

Low maintenance grassland, intended to blend with surrounding vegetation, should receive spring and autumn cuts only. This allows the site to retain a 'natural' appearance without causing an accumulation of dead vegetation which may limit further growth. In general, all newly established grassland should receive a preliminary cut as soon as possible to ensure that weeds do not become established and to allow closing of the sward.

10.10 Wild species

Modern agricultural practice and urban expansion have reduced the area of wilderness in the region of most large cities. Redevelopment of land therefore can provide opportunities to reinstate various wildlife habitats in urban areas. One such method of developing this type of landscape is the use of seed mixtures containing native and wildlife species. When choosing a seed mixture it is essential that the plants are compatible with the location of the site. The input of nutrients must also be managed carefully. If fertilizing is too heavy, then fast growing grasses may dominate the wild species. A small dressing of slow release compound fertilizer should be applied in early spring during the first years of the establishment phase. Wild species are rarely able to tolerate hard cutting and a single mowing in early autumn is usually adequate to maintain the site in a tidy condition. Areas of wilderness require very little maintenance and this treatment is most suitable for small sites which are inaccessible to earth moving or agricultural machinery. A further advantage is that wild species can generally be established on very poor substrates to produce a visually attractive cover (ref 5.)

10.11 Trees and shrubs

Where trees and shrubs are to be planted as part of a landscaping scheme, the species should be chosen to be compatible with the location. For this reason, natural associations of common British trees are commonly planted and are often the most visually attractive. By planning a careful balance of species, the rate of natural succession can be increased, producing a vigorous vegetation which requires little maintenance.

The success of any tree planting scheme heavily depends on the quality of the plant material. Because of the poor condition of some reclaimed sites, it is inadvisable to attempt to establish trees larger than half standards. It is usually more suitable to plant 'whips' because they are less costly to replace, and are less prone to fatal damage during transportation and planting. In general, 'whips' should be planted in slits, notches or pits. For all methods, the size of the cavity should be large enough to contain the roots without causing any damage. Pits should normally be no smaller than 300 mm diameter by 300 mm deep, but of course larger trees do require larger pits. When the pits are backfilled with a suitable soil/organic matter mixture, the surfaces should be left lower than that of the surrounding soil, so that surface water drains into the pit. Watering is often essential for successful establishment of trees and shrubs on all sites. The frequency of watering must be related to the season and weather conditions, and must be adjusted as tree or shrub growth increases.

10.12 Timing

Grass seed should be sown in the early spring or late summer and early autumn. This gives the seed time to germinate and establish before the onset of adverse weather conditions, that is, dry, hot summers or cold, wet winters.

For amenity grassland, subject to heavy use in the summer, it is preferable to sow in late summer or early autumn. For most other uses, however, a spring sowing is often the best.

Bare rooted trees and shrubs should be planted throughout the winter period when the tree or shrub is dormant. The more active the tree or shrub is in terms of growth, the more likely it is to die from transplant shock.

Trees or shrubs in pots or bound with soil around their roots can be transplanted at any time of the year, but may have to be watered during dry summer months.

10.13 Plant and soil interactions

In developing any landscape scheme on a contaminated land site reclaimed by methods other than complete removal of the contaminants, the ability of plant roots to interact with covering systems and the contaminants themselves must be considered.

It is particularly important to prevent plant roots from coming into contact with contaminants isolated, encapsulated or covered by the reclamation scheme. Transport of contaminants from the root region underground to the vegetative portions of plants above ground can provide a very effective means of pollutant transfer when contaminants are not present in phytotoxic forms or quantities. When contaminants are phytotoxic, penetration by root systems will, of course, result in death of the vegetation, and the failure of the landscape scheme.

The most obvious method or preventing roots from coming into contact with contaminants is to increase the depth of covering materials, and by careful choice of shallow rooted species in the landscape design. However carefully this is carried out, so called shallow rooting species may penetrate to surprising depths. For example, in dry summers the roots of a grass sward may penetrate beyond 1200 mm. In moist summers grass roots may still penetrate to 500 mm and this can often be deeper than cover layers applied in many reclamation schemes. Other preventative methods of inhibiting root growth into contaminants must therefore be considered.

The materials which prevent root penetration most effectively are usually coarse-textured with many air filled voids. Such materials include brick and other demolition rubbles and gravels. Roots which penetrate into the voids are then killed by desiccation. Coarse layers of this type fail to prevent root penetration when fine material from overlying layers fill up the voids, thus providing channels for root growth. This may be prevented by incorporating a blinding layer of less-coarse material between the root barrier and the soil or by gradually reducing the particle size of the barrier with distance from the waste.

Barriers which prevent root development again limit the ability of the vegetation to withstand drought. If such a barrier is used, the materials above the barrier should again be capable of retaining sufficient water to maintain plant growth during periods of drought.

10.14 Conclusions

The fact that the nature of contaminated land is diverse ensures that a number of reclamation or treatment options are available. If landscaping design is considered as part of the reclamation scheme, then it can take various forms depending on final land use and the way in which a site has been reclaimed. In any event all landscape schemes must consider the requirements of plants for maintenance of good growth, but at the same time ensuring that the complex reaction of plant roots themselves, plus soil structure and chemistry are taken into account.

References

1. Dutton, R.A. and Bradshaw, A.D. (1982) Land reclamation in cities. HMSO, London.
2. Department of the Environment (1987). Joint Agricultural Land Restoration Experiments London, Department of the Environment.

3. Townsend, W.N. and Gilham, E.W.F. (1975). Pulverised fuel ash as a medium for plant growth. In: *The Ecology of Resource Degradation and Renewal.* M.J. Chadwick and G.T. Goodman, (eds.) Blackwell, Oxford, 287–304.
4. Bradshaw, A.D. and Chadwick, M.J. (1980) *The Restoration of Land. The Ecology and Reclamation of Derelict and Degraded Land.* Studies in Ecology, Vol. 6, Blackwell, Oxford.
5. Cole, L. and Keen, C. (1976). Dutch techniques for the establishment of natural plant communities in urban areas. *Landscape Design*, **116**, 31–34.

11 Hazards from methane (and carbon dioxide)

D.L. BARRY

The principal gases generated in domestic landfill, or similar sites with degradable organic material, are methane and carbon dioxide. The latter is dominant in the early stages of material decomposition while conditions are aerobic. As conditions become anaerobic due to the gradual exclusion of air and the conversion of oxygen in the original voids, methane becomes the more dominant. In the anaerobic phase, the general range of methane and carbon dioxide proportions are 50–70% and 30–50% respectively. Both these gases can present considerable hazards for building developments on or near landfill sites.

The fundamental distinction to be drawn between landfill gases and other ground contamination issues is in respect of the relative ease with which potentially hazardous conditions can arise, and the potential consequences of an explosion especially in a building. Accordingly, a high degree of caution must be exercised when dealing with this gas, notwithstanding our understanding of the explicit hazards involved.

Any development on a gas-active site involving buildings in which hazardous conditions could arise, must be given most serious consideration and be backed by careful investigation and design of control measures. It is the DoE's clear policy that no such development should occur on landfill sites until the gas generation has virtually or finally ceased.

Considerable difficulties can be experienced in measuring and assessing the actual gaseous conditions on a site. A selection of investigation techniques and instruments are readily available but care must be taken to ensure that these are used and interpreted appropriately. Elevated gas readings are more easily interpreted than low or zero readings since the latter can beg many questions about the adequacy of the investigation and the residual gas generation potential.

Measures can be applied to prevent or control gas movements within or from the site and these include barrier systems, passive venting systems and pumping systems. The effectiveness of any one or combination of these will depend on the proper definition of the risks in the first instance, the proper installation of appropriate measures, and on proper monitoring and maintenance. The latter aspects must be given greater than normal emphasis due to the general physical instability of the ground mass and hence the potential for differential settlement and its consequential impact on control measures, such as barriers and slabs.

11.1 Introduction

Two of the most significant differences between gases and other chemicals encountered on contaminated sites are that there may be little warning of hazards, and the consequences of gas hazards can be disastrous. Landfill sites are well known for their production of methane, in particular, and, while there might not be any warning of methane *hazards* on a particular site, there is more than ample warning of the *risks* from developing on or near to such sites.

Wherever degradable organic matter exists, microbiological reactions are likely, usually resulting in gases being generated. The principal microbiological reactions under consideration here are generally classified as anaerobic fermentation and in domestic waste landfills anaerobic conditions can arise as waste compaction takes place. This compaction restricts the ingress of air, and the oxygen from any air present is gradually used up during the preliminary aerobic stages of decomposition. Also, the gases being generated will tend to

Table 11.1 Typical or possible occurrences of hazardous gases on contaminated and other sites

Site or ground type or feature	*Principal gases*
Refuse disposal sites*	$CO_2 + CH_4 (+H_2 + H_2S$ *et al.*)
'Inert' refuse	H_2S
Peaty/marsh/silty ground e.g. riverside, dockside, coastal	$CO_2 + CH_4$
Coal mines/seams (or fissures connected thereto)	CO_2 (black damp) or CH_4 (fire damp)
Action of acids on carbonate rock	CO_2
Industrial refuse	Various organic vapours
Gasworks and coking plants	H_2S; HCN
Gas mains/fuel lines/tanks	Various hydrocarbons
Sites polluted with cyanides	HCN
Sites where anaerobic microbial action occurs on (i) sulphur-containing organic materials, (ii) sulphates, (e.g. gypsum) (iii) sulphides	H_2S
Sites where sulphides are oxidized on exposure to air	SO_x
Slow oxidation of buried coal waste	Gas mixtures rich in CO_2, CO and CH_4 and deplete in O_2
'Oil' spillages	hydrogen, acetylene, propane, butane, petroleum vapour
Oil refineries	H_2S + various hydrocarbons

* Many 'non-domestic' sites containing paper, wood and other carbon-based materials can also produce considerable quantities of gas, under the right conditions.

prevent the slow natural ingress of air and hence reactions eventually take place in an environment totally deficient in oxygen.

The range of circumstances in which such gases are generated is wide, ranging from natural conditions such as carboniferous rock formations under acidic conditions, and peaty or marshy conditions, to man-made conditions such as waste disposal sites. These gases, which are usually dominated by methane and/or carbon dioxide, can have a profound influence on the choice of use or after-use that such sites offer. On the one hand there can be significant hazards to buildings and their occupiers while, on the other hand, plant life can be seriously affected.

The most common type of site with active gas generation is waste disposal sites, especially those with domestic or readily degradable refuse. These sites can appear very active (e.g. bubbles in water pools) or be dormant for many years either during or on completion of disposal operations.

In this chapter, emphasis is placed on gases from domestic refuse or similar sites (as opposed to toxic or industrial waste sites) and so the problems of volatile organic vapours, for example, are not addressed. (Table 11.1 gives an indication of some of the gases that can be encountered on a range of contaminated sites or other ground conditions on which development is frequently considered.)

11.2 Relevant gases and their principal characteristics

11.2.1 *Introduction*

'Methane' is frequently used to describe the range of gases that are produced in domestic refuse and similar landfill sites. However, there are a number of other gases that can be generated which could be hazardous during or after site redevelopment. These gases include carbon dioxide (CO_2), carbon monoxide (CO), and hydrogen sulphide (H_2S). These gases can be readily released into the atmosphere (whether through buoyancy effects or pressure gradient), and are even more likely to exist in trenches and excavations during construction periods. The presence of these gases (individually or collectively) can also result in oxygen deficiency which is a potentially critical condition in its own right and one which is not always recognized as such.

11.2.2 *Typical landfill gas composition*

While landfill gas is considered to consist 'typically' of two-thirds methane and one-third carbon dioxide, the actual percentages of these, and the range of other gases, can vary widely. The gas composition also changes with the life of the tip, particularly at the early and later stages (Figure 11.1), and also the composition can vary with the forced withdrawal of gas by pumping.

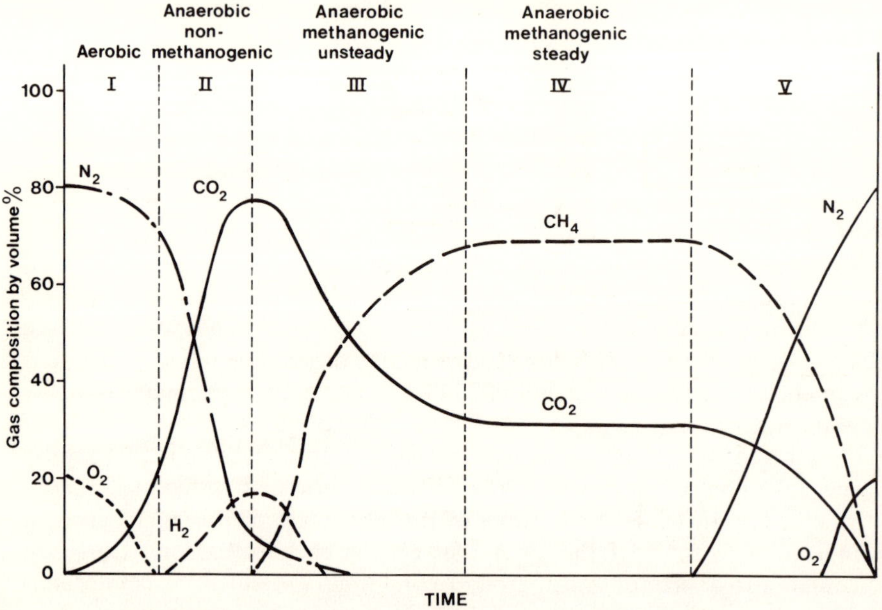

Figure 11.1 Phases in landfill gas generation (schematic)

11.2.3 *Landfill gas generation—principal factors*

The generation of gas from the microbiological action in landfill sites is complex (Figure 11.2a) and depends on many factors that affect both the rate of generation and its duration. The principal factors are shown in Figure 11.2b. In general, any waste material susceptible to aerobic degradation can also be degraded anaerobically, albeit usually at a reduced rate. This rate can also be influenced by the mixing of wastes e.g. sewage sludge mixed with refuse will enhance methane generation, while certain industrial wastes can inhibit it.

The decomposition of organic wastes is postulated to occur over five stages, as follows (Figure 11.1).

Stage I Aerobic decomposition in which entrained oxygen is converted to CO_2. Heavy compaction, good daily cover and rapid infill to deep layers can reduce the importance of this stage. This stage has normally finished before reclamation starts but may recur if the refuse is exhumed during construction work.

Stage II Anaerobic decomposition commences as the oxygen becomes depleted. CO_2, and even hydrogen, is produced and the nitrogen from the air is progressively displaced.

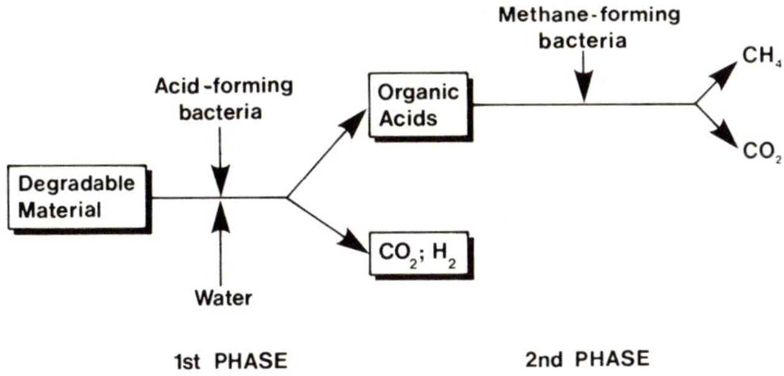

a) Simplified decomposition process

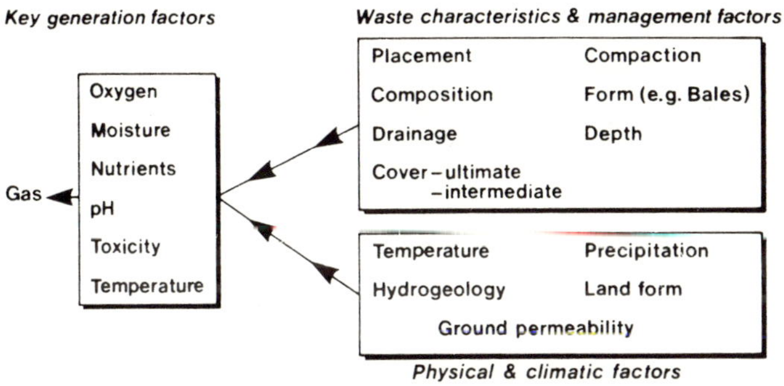

b) Factors affecting gas generation

Figure 11.2 Landfill gas generation

Stage III Anaerobic methane production rises to a peak and carbon dioxide concentration falls. Nitrogen concentrations drop to zero.

Stage IV A relatively steady methane production rate can occur over many years depending on the physical and chemical factors and the extent to which gases can escape either through the top or into surrounding ground. Gas compositions at this stage are usually in the ranges, methane 50–70% and carbon dioxide 30–50%.

Stage V A steady decline in methane production and a corresponding increase in aerobic conditions.

It can be misleading to suggest time scales for each of these stages since very wide ranges have been observed, largely due to the variations in critical factors (Figure 11.2b). For example, Stage II has been known to have commenced in a matter of months, while Stage V might not be reached for many decades. There can be large volumetric peaks and troughs in the various stages due to changing conditions such as moisture content or temperature, for example. The only generalization that can be made about landfill gas is that gas can continue to be generated (albeit not necessarily in critical volumes) until all relevant material has been degraded.

11.3 Gas flow in and from landfill sites

11.3.1 *Introduction*

Gases flow through refuse or soils either by convection or by diffusion. Convection occurs when total gas pressure is not uniform throughout the system, i.e. when a total pressure gradient exists and the gases move from high pressure areas to low pressure areas. Diffusion occurs when gas concentrations in the landfill are not uniform and the gas flows from areas of high concentrations to low, i.e. down the partial pressure gradient.

Both types of flow are particularly important in assessing the potential lateral migration of gas off-site and the effectiveness of passive control systems. Where gases are withdrawn by pumping, convection flows are more significant.

11.3.2 *Buoyancy effects*

Although methane itself is lighter than air, when a landfill gas mixture escapes from the ground it may not be very buoyant unless, of course, there is a considerable temperature differential (see Table 11.2). Thus, the buoyancy of a 5% methane/5% CO_2/90% air mixture is almost the same as 100% air. In simple terms, therefore, low emission rates are less likely to be buoyant than high emission rates because, apart from any 'kinetic' benefit, the temperature differential for high rates is likely to be greater (i.e. less opportunity to cool down). Although usually generated at temperatures well above ambient UK

Table 11.2 Gas densities* of air and landfill gas mixtures

		Moisture content		
Gas mixture	*Temp.°C*	0%	50% *Saturation*	100% *Saturation*
'Normal' air	15	1.226	1.219	1.217
70%CH_4/30%CO_2	15	1.038	1.029	1.020
	25	1.003	0.987	0.971
	35	0.971	0.943	0.918
	45	0.940	0.898	0.857
50%CH_4/50%CO_2	15	1.276	1.266	1.255
	45	1.156	1.104	1.052
5%CH_4;5%CO_2;90% air	15	1.231	1.220	1.210

*g/l at 760 mm Hg

temperatures, landfill gases pass through material near the surface which is at or near ambient air temperatures. Accordingly, depending on flow velocity, the emerging gases may not differ greatly in temperature from the ambient value.

11.3.3 *Gas concentrations*

During the 'unsteady' phases of landfill gas production the source concentrations of the major gases can vary considerably (Figure 11.1). However, when a methane concentration is measured at, say, 1000 ppm, it must be remembered that at the origin of that gas source, the concentration may have been up to 70% (assuming that CO_2 will account largely for the balance). Thus, somewhere en route the gas may have been diluted by air or other gases.

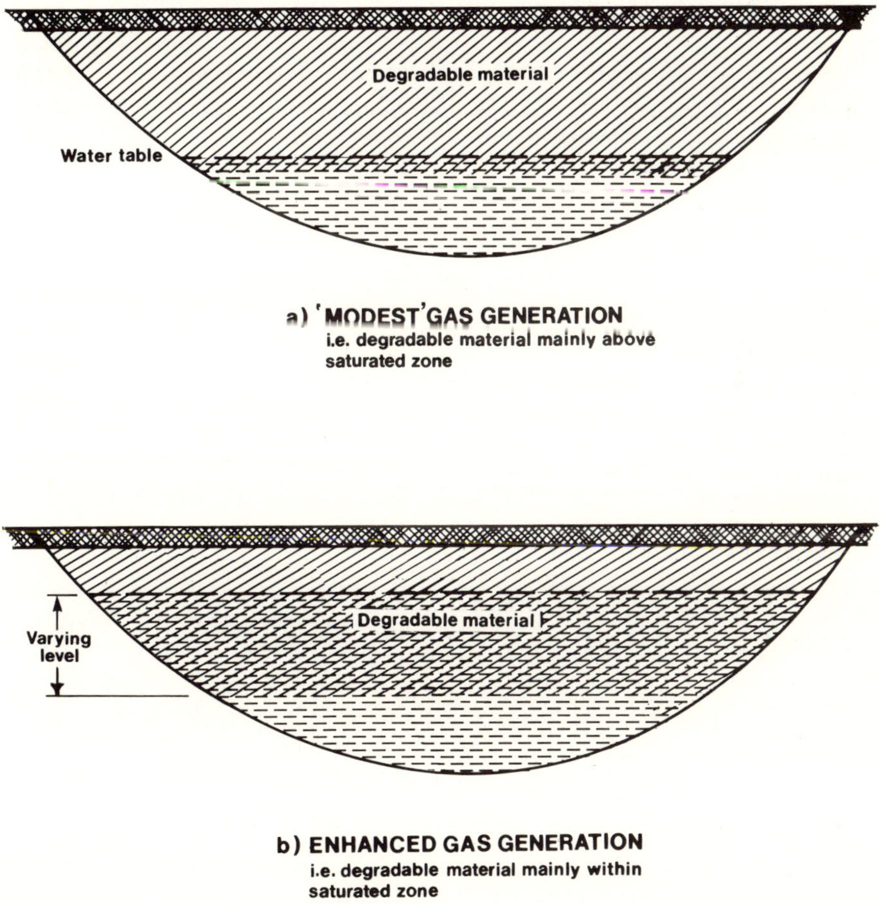

Figure 11.3 Influence of water table on gas generation

This dilution would normally occur while the gas was travelling through the aerated soils and the rate of dilution should be much greater at low gas velocities. Similarly, if the emission rate from a fissure were increased considerably (due to a rise in water table level, for example) the emission concentration might also increase to a corresponding degree.

The proportion of CO_2 in landfill gas can vary at the point of generation as well as subsequently if the gases should pass through water, for example. Since methane is only slightly soluble in water, and CO_2 is soluble both in water (forming carbonic acid, H_2CO_3) and in most organic liquids, the passage through water and leachates can increase the methane proportion. On the other land, methane can be oxidized by certain soil bacteria producing CO_2. Thus the concentrations of gas constituents at some distance below the

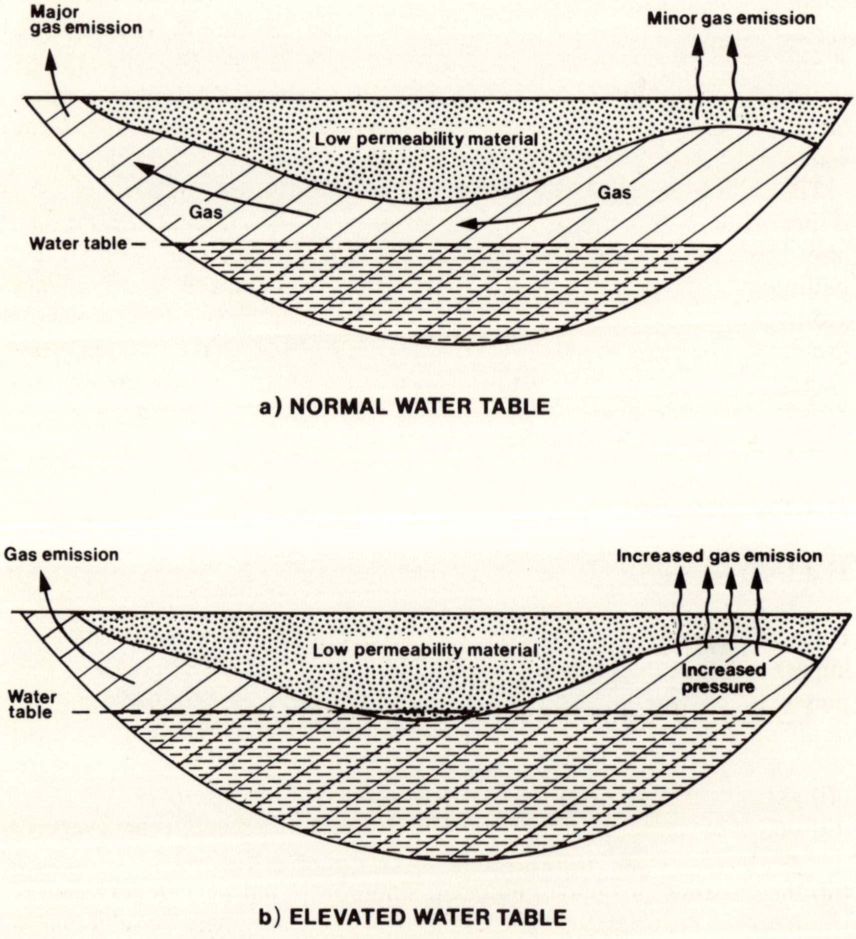

Figure 11.4 Effect of water table on gas emissions

ground surface could be altered significantly by the time they emerge at the surface, through 'secondary' physico-chemical and biological processes.

11.3.4 *Effects of water table*

The effects of the groundwater table and its mobility can be manifest in a combination of physical and chemical forms. These effects can be additive. Optimum gas generation conditions are considered to occur (other things being equal) when the material is more than 50% saturated (1). In normal ground these conditions occur above the groundwater table (i.e. in the semi-saturated zones), and the thickness of these zones is dependent on the soil structure e.g. in coarse gravels there is little capillary effect above the water table, while in clayey soils the effect is very considerable. In the heterogeneous conditions of a typical landfill the saturated zone area is not predictable, but it is reasonable to assume that optimum conditions are more likely to occur in the region close to the main water table or perched water tables. Thus changes in the level of the water table can cause different materials to be more or less 'active' (Figure 11.3). Moreover, water can be retained by the fill when the water table drops or it can become saturated by surface water ingress.

The physical effect of a rising water table level can be very pronounced. This is simply due to the water pressurizing the gas volume. A rise in water level can also affect the location at which gas emerges, for example, the 'normal' pathway may be cut off (see Figure 11.4). Thus, seasonal effects such as hard ground frost and changes in atmospheric pressure, or other effects, such as groundwater abstraction, can significantly change the rate and location of gas emissions. This highlights the fundamental problem in interpreting gas emission data taken over a short period.

11.4 Gas measurement

11.4.1 *Introduction*

Before a solution to a problem can be developed, it is of fundamental importance that the problem is first defined as accurately as possible. This presents considerable difficulties on landfill sites since gas measurements can be difficult not only to take, but also to interpret. This is because:

(i) gas *emission* rates can be confused with gas *generation* rates
(ii) the act of measuring, for example by sinking a borehole, creates a vent for gas release, and
(iii) the 'existing' or pre-development condition of the site can vary greatly from the post-development condition, e.g. the site might be subjected to dynamic consolidation, the water regime varied, or the site capped.

Therefore, it is important to recognize the influence of the act of measuring in the same way that, in geotechnical engineering, true 'undisturbed' samples are 'impossible' to take but, nonetheless, are essential to the design process.

A particularly relevant publication dealing with current methods of gas measurement and hazard assessment has been produced (1985) by the Greater London Council (2).

11.4.2 *Gases to be tested*

There appears to be a widespread tendency in landfill gas sampling to test only for methane. This could prove a critical restriction since the concentrations of carbon dioxide could, in some circumstances, be more significant. For example, carbon dioxide at 4% concentration could present a far greater hazard to personnel than methane at the same concentration. (The 10 minute occupational exposure level (OEL) for CO_2 is 15 000 ppm, i.e. 1.5% (3). Even if it is clear that the only concern about landfill gas is in regard to flammability, then carbon dioxide concentrations should still be tested. This is because CO_2 may act as a precursor for methane (i.e. Phase I) or it may indicate the oxidation of methane. Elevated CO_2 levels can, therefore, sometimes suggest that methane is being generated but not emerging as such or that the generation phase is still largely aerobic.

11.4.3 *Sampling and monitoring systems*

11.4.3.1 *Ground probes*. The simplest method of monitoring gas composition in soils or refuse is by sinking a hollow rigid pipe into the ground, with a tube connected to a gas detection device. A hole can be made for the probe with a steel rod, and after inserting the probe the hole is sealed at the surface. Using an aspirator a gas sample is then drawn from the probe. Gas samples should be taken immediately after forming the probe hole and as frequently thereafter as suggested by the initial readings, i.e. check if the readings are steady or fluctuating.

This method may often give a positive gas reading, but failure to do so does not necessarily mean absence of gas on the site or in that general location. This is because vertical gas migration in an area might be prevented or reduced by an underlying low permeability layer. Moreover, methane generation is usually low or non-existent at shallow depths because of the oxygen presence. Thus, such probes may often fail to detect the high concentrations that may exist at a greater depth.

Although results from such probes are not repeatable, i.e. a new hole must be made for each survey, this method can be a very useful and cheap preliminary test for the presence of gas.

11.4.3.2 *Permanent or deep monitoring.* A permanent or deep monitoring probe may require that a hole is mechanically drilled or driven into the ground using conventional geotechnical ground investigation techniques. Using those techniques, a gas probe can be inserted to the required depth, the relevant zone depth backfilled with pea gravel (or similar) and a concrete seal placed at the top of the probe hole (Figure 11.5). Probes usually consist of perforated tubing.

If samples are required from different zones in the same area, it is usual to install either multiple boreholes or to set a number of probes within one borehole. These probes reach to different zones which are separated by seals to prevent gas migration between zones.

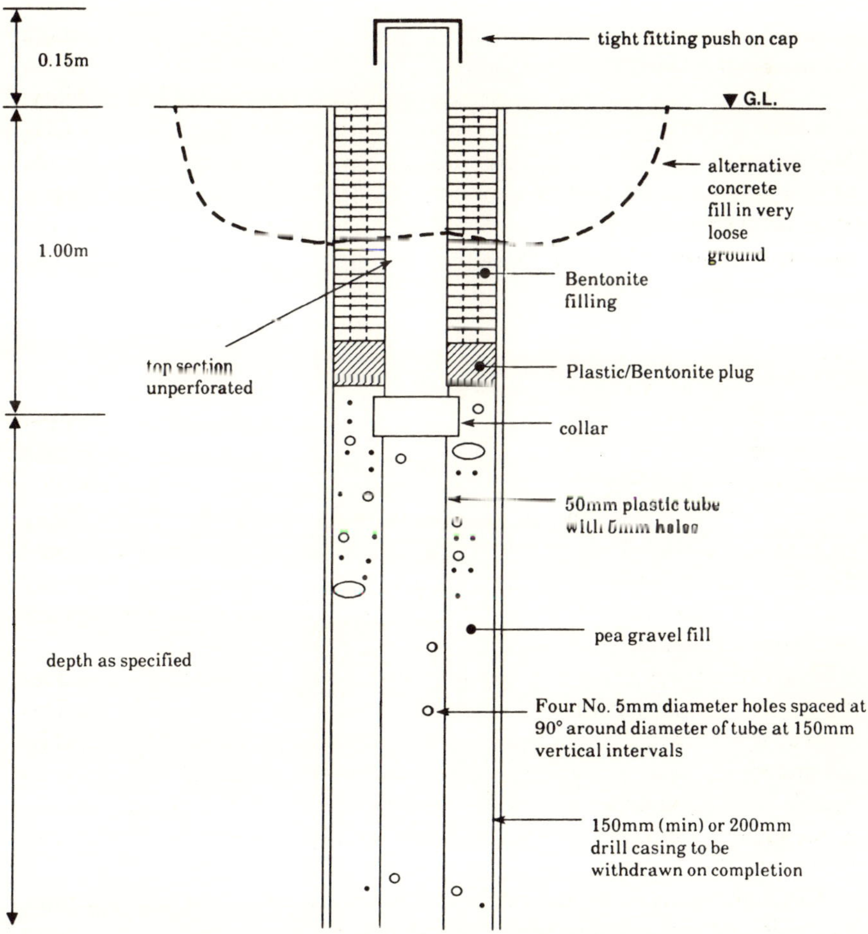

Figure 11.5 Typical methane probe construction (after ref. 2)

11.4.3.3 *Trial pits*. One of the simplest, and reasonably effective, gas survey techniques is to dig a trial pit (to say 5 m) and insert the perforated tubing before backfilling. This approach has a number of advantages; for example it affords the opportunity of examining the nature of fill materials. The method is quick and relatively cheap. Extreme caution must be exercised in trial pitting landfill sites (as with all potentially contaminated sites), not least because the ground mass is usually very unstable (4).

Probes for these pits should be robust enough to withstand the stresses caused during backfilling, in particular, when sharp or heavy objects could be placed against them. A further problem could be from the blocking of perforations during backfilling. It is difficult to resolve this without resorting to a more carefully controlled backfilling operation whereby granular material was progressively placed around the pipe. This approach could largely negate the cost and time benefits of the system compared to boreholes. Probes should be capped off and protected and the probe hole sealed for the top metre or so to prevent gas escape.

It can be argued that trial pits have an excessive effect on the conditions being monitored as they allow air (and perhaps water) into the ground mass and also the backfilling changes the state of ground compaction. However, this can be a relatively short-term impact and the principal short-coming can be when low gas concentrations exist. However, if high concentrations are detectable shortly after probe installation and backfilling then that clearly shows that gas is 'available' for release, albeit through a disturbed zone.

11.4.4 *Taking gas measurements*

There is, as yet, no approved or 'preferred' on-site method of measuring landfill gas. Moreover, since the emphasis is frequently on methane only, the presence and potential influence of the other gases is not always appreciated in terms of the accuracy of monitoring devices. Variations between site and laboratory measurements can sometimes be explained by such influences. Gas chromatography is considered the only 'sure' method of determining the composition of a gas sample, but for multiple samples this can become very expensive, and may not be practical or necessary in a majority of instances. Portable gas chromatographs are, however, available in the UK.

Care must be taken in the choice of site measuring instrument to ensure that the gas conditions fall within the range specified by the manufacturer. Some of the current popular portable measurement devices were developed for the gas industry (for leak detection, for example). All these are not yet fully proven in the range of conditions encountered in landfill sites where gases are usually saturated and have a considerable proportion of CO_2, two factors which clearly distinguish the gas from natural gas. For example, some instruments are considered to be satisfactory for low methane concentrations provided the oxygen level does not fall below 10%. Since such instruments work on a 'hot wire' principle there is also a slight risk of ignition, which could be critical in some instances.

If a meter shows a '% LEL*' range, then one must be careful about interpreting a reading of say '50%' since it can be readily interpreted as representing '50% of the LEL of methane in air'. This is not the case and what it actually represents is the 'explosiveness' of the gas mixture being measured. Thus the meter is a 'hazard detector' rather than an explicit methane detector. In practice, however, this could often be an academic difference but it is important to recognize that wrong interpretations can be made.

A frequent method of testing for gas in different zones is to carry out the tests during the drilling of the borehole, that is, by sampling from the bottom of the borehole before the steel casing is driven into that zone.

Measurements at successive depths can indicate the variations in gas migration or generation zones. On completion of the borehole, gas emission tests are usually carried out for the hole as a total, with no distinction being drawn between different zones of origin. These initial measurements, however, cannot distinguish between generation rates and 'emission' rates in a particular area, since the probe may have tapped a large gas reservoir resulting in misleading information. Emission rates of gas are obviously related to the gas pressure, which is a function of both generation rates and ease of migration. Measuring such rates are usually necessary for an assessment to be made of the volumes of gas that could emerge if, for example, foundation piles were driven through the site (i.e. allowing gas to escape around the pile perimeter) or that might emerge through ground fissures.

11.4.5 *Interpreting gas measurements*

Notwithstanding the technical problems of properly taking gas samples, there are considerable problems in interpreting the resultant data. For example, how do the collected data relate to the wider conditions within the tip, both now and in the future? High concentrations and large volumes unquestionably require active control and monitoring measures, whether involving gas pumping and/or forced ventilation. Without under-estimating the difficulty in designing and maintaining such controls and precautions, this situation is probably more clear cut than the low volume/concentration situation where the need for any measures is not always explicit. This is related to the fact that control measures can be difficult to scale down and so the cost penalty for dealing with low volumes and concentrations can be similar to that for greater ones. In simple terms, high methane measurements are often easier to interpret than low ones since there is little argument about the possible consequences of explosive concentrations.

It must be considered that any landfill site with degradable organic material within its mass is capable of producing methane. Since critical gas migration is

*LEL = Lower Explosive Limit

mainly through convection, it is not normally possible to get high volumes of low methane concentrations (except, of course, when the balance of the gas is CO_2 as can occur in the early stages of decomposition). Therefore, if zero or very low concentrations are consistently measured over a number of seasons the following questions must be posed before the assumption is made that the methane generation stage has been completed.

Is the methane migrating elsewhere? A low permeability layer could be preventing upward migration.
Have probes been sunk to the base of the tip?
Have solid samples been analyzed for degradable content?
Is there a low pH value or chemical 'poison' causing a suppression of methanogenic bacteria?
Is the tip deficient in moisture?
What CO_2 levels have been recorded?
Is the tip poorly insulated, causing aerobic conditions and not permitting the ground mass temperature to rise (more typical of above-ground fills).

11.5 Gas control measures

11.5.1 *Introduction*

The need to control gas may arise from two requirements: to prevent migration off-site to an inherently safe area, and to enable the gas-producing site to be developed safely. The discussions below of options for dealing with the latter must not be interpreted as supporting the development of such sites. Indeed, it is the view of ICRCL and other DoE bodies that development on any site known or suspected to produce methane should not take place before emissions have ceased. However on the basis that some remedial action will be necessary if development takes place, then a number of options exist for preventing or controlling both surface emissions and lateral migration off-site. These options can, in simple terms, consist of barrier systems, venting systems, pumping systems or a combination of these.

11.5.2 *Options for reducing hazards to buildings*

11.5.2.1 *Introduction.* There are two basic options for reducing or eliminating methane hazards in buildings:

(i) Preventing gas from reaching the building, and
(ii) Providing adequate ventilation in spaces to which gas can or may migrate in or near a building.

In practice, a combination of these two options may be utilized but, unquestionably, the former option should be the primary choice. A further option might also be presented, namely, to avoid confined or potentially confined spaces, where *any* methane is known to exist. The ultimate option is to avoid constructing buildings on gas generating areas, or on areas to which gas could reasonably be expected to migrate.

11.5.2.2 *Building design.* The broad principle generally adopted in trying to prevent methane entering buildings, is to construct the building on a base that is 'impermeable' to gas penetration. The potential weakness of this concept is that the relevant slabs may not be or remain fully intact, e.g. service ducts or ground settlement can fundamentally alter the integrity of the base *vis à vis* the passage of gas. While some of these discontinuities may well be essential to the building requirements, they are most likely to present the most susceptible routes for gas ingress. Taking services up the outside of a building to above ground level before entering the structure can considerably reduce this risk since any gas which follows the underground pipe or duct will undoubtedly become diluted as it emerges from the ground. The benefit of this system could be largely negated if the entry point were sheltered by a projection for example (Figure 11.6). Similarly, large gas volume emissions could become trapped by overhanging eaves, for example, and so enter a roof space. The risk of this occurring must be considered low, however, unless the development was on a very active site, the external wall face was free of projections (which create air turbulence) and the weather conditions were still. It is also possible that methane could enter via an air brick or similar. This possibility can be eliminated with reasonable certainty by extending the floor slab (Figure 11.7).

11.5.2.3 *Ventilation of underfloor voids.* Elevating a floor slab above ground level and creating a vented void beneath, should reduce the possibility of critical gas concentrations entering the building. The effectiveness of this system is, of course, influenced by two other factors, namely whether or not critical concentrations can enter the void space in the first instance, and whether a pathway into the building does actually exist. The first of these two conditions cannot always be predicted with certainty since:

(i) Gas generation may be enhanced after development if the principal parameters discussed earlier are critically altered, or
(ii) Gas migration patterns in the ground mass may be altered, resulting in critical concentrations and volumes emerging under the floor slab, or
(iii) Predevelopment measurements may not have properly reflected either the actual or potential gas regimes, e.g. seasonal variations or local hot spots.

The provision of a gravel filter blanket beneath a void space (or any floor slab) has the potential for reducing peak gas concentrations that emerge from the landfill. Thus the gas is likely to be diffused and hence diluted, as it travels through the filter. Such a design together with a special service duct system has been installed at a school in Devon (Figure 11.8).

11.5.3 *Migration barriers*

In practice few, if any, barriers have zero gas permeability but many have permeability values sufficiently low that they can prevent the migration of critical volumes of gas. Cut-off barrier systems can be either horizontal or

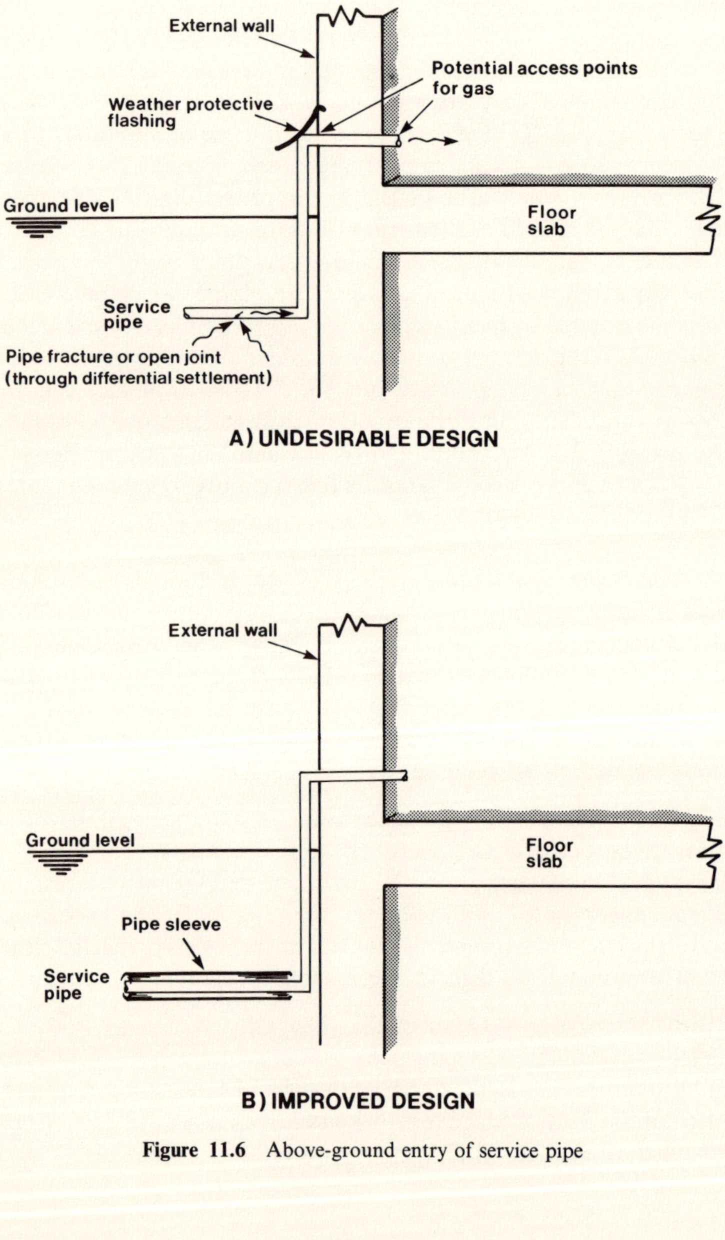

Figure 11.6 Above-ground entry of service pipe

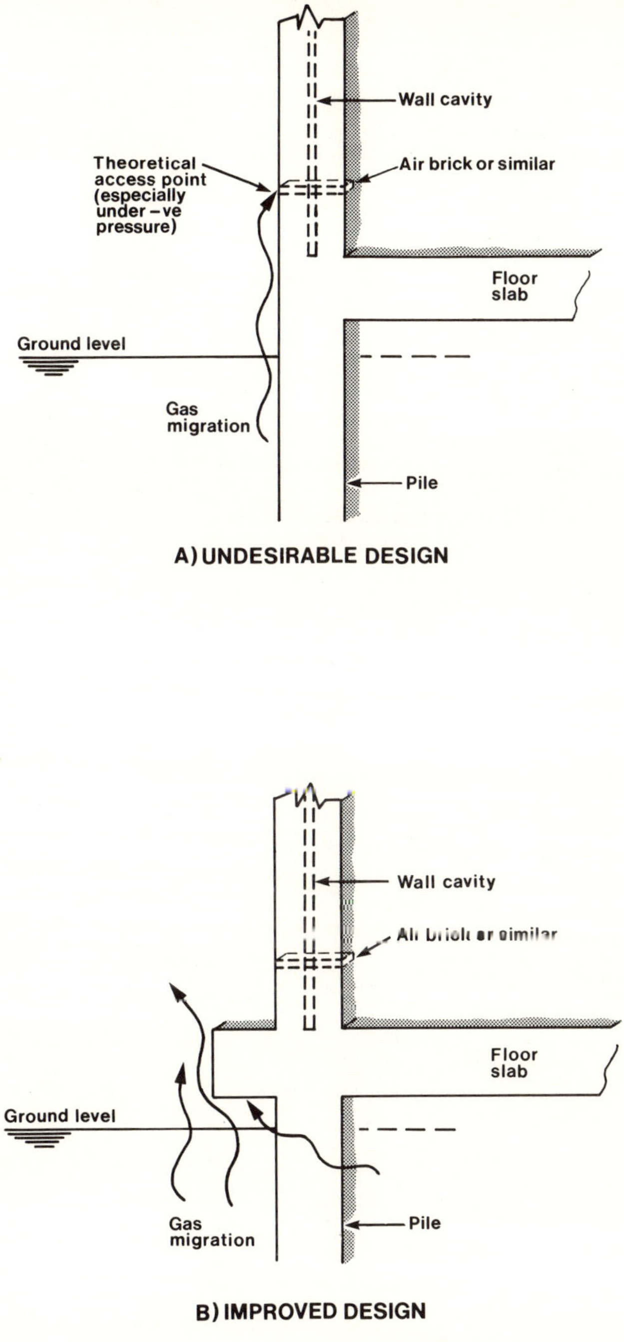

Figure 11.7 Gas entry through external wall

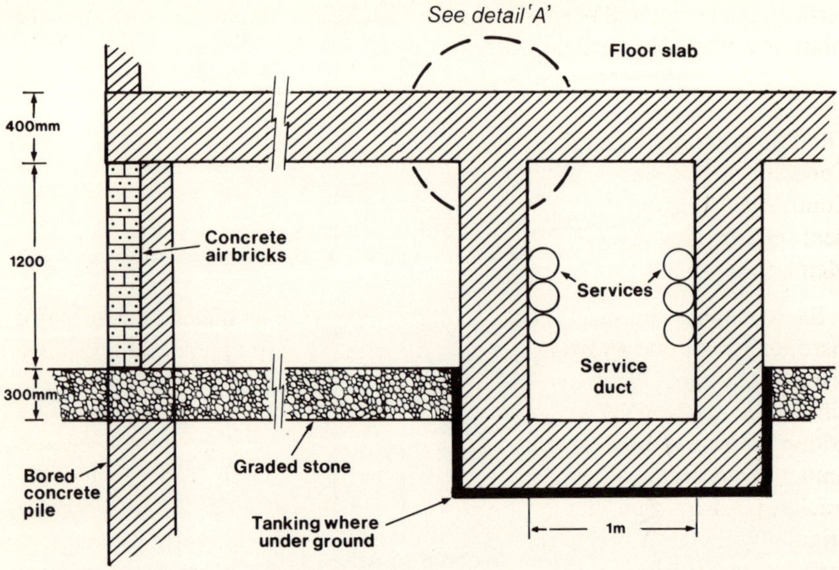

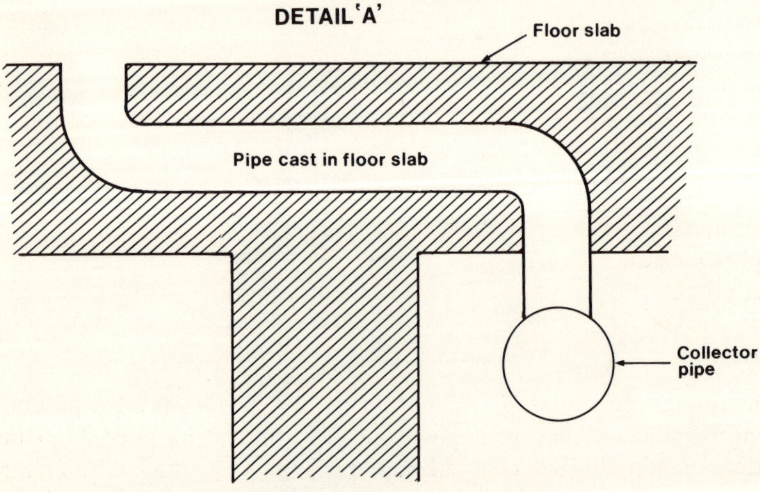

Figure 11.8 Example of floor slab design

vertical, the horizontal barrier being aimed at preventing or reducing upward migration while the vertical prevents or reduces lateral migration.

Horizontal barriers
The following are examples:
Concrete raft or slab
Compacted hoggin
Bentonite or clay layer
Membrane.

Each of these can meet a design objective as long as their integrity is not disrupted, with due allowance for quality control during construction. This disruption is most likely to result from settlement of the ground (particularly, differential settlement) and/or from 'mechanical' interference through, for example, sewers, ducts or piles penetrating the barrier. Clay, bentonite and similar layers on the other hand can be affected additionally by drying out and cracking. Also when re-moistened, clays do not expand to 100% of the original volume. This is a particular problem when constructed close to the surface and is further exacerbated by 'long hot summers'. Membranes can also deteriorate through chemical attack. In essence, therefore, the integrity of horizontal barriers, of virtually any construction, is liable to serious medium to long-term degradation on many landfill sites. These barriers, however, are not always required for absolute control of gas flow but more to exert a major migration influence during abstraction. In addition, physical weaknesses that develop might be modest in scale and effect and may be capable of restoration. Any deterioration in the integrity of the barrier, such as those described above, can however negate the fundamental objective of the barrier if gas passes through the barrier and enters a critical location or region above.

Vertical barriers
The following are examples:

Grout curtain
Diaphragm wall
Sheet piling
Slurry trench
Vent trench.

The primary functions of such barriers within or outside a gas generating site are similar, namely to prevent gas from migrating laterally. Thus a building on a landfill site could be isolated from the ground surrounding it, and in this way any venting system has to cope merely with gas generated beneath the building.

A further potential benefit of such a technique (leaving aside the cost implications) is that the gas generation rate within the confined space could be enhanced and so reduce the active life of the biological action. In any event, the long term durability of such a perimeter barrier must be taken into account.

The major practical problem with vertical barriers within rather than adjacent to a site is in their installation, and the scale of these problems is fundamentally greater than when barriers are installed outside the filled area. All the above mentioned examples of barriers are installed frequently for engineering purposes and so considerable relevant experience exists. The

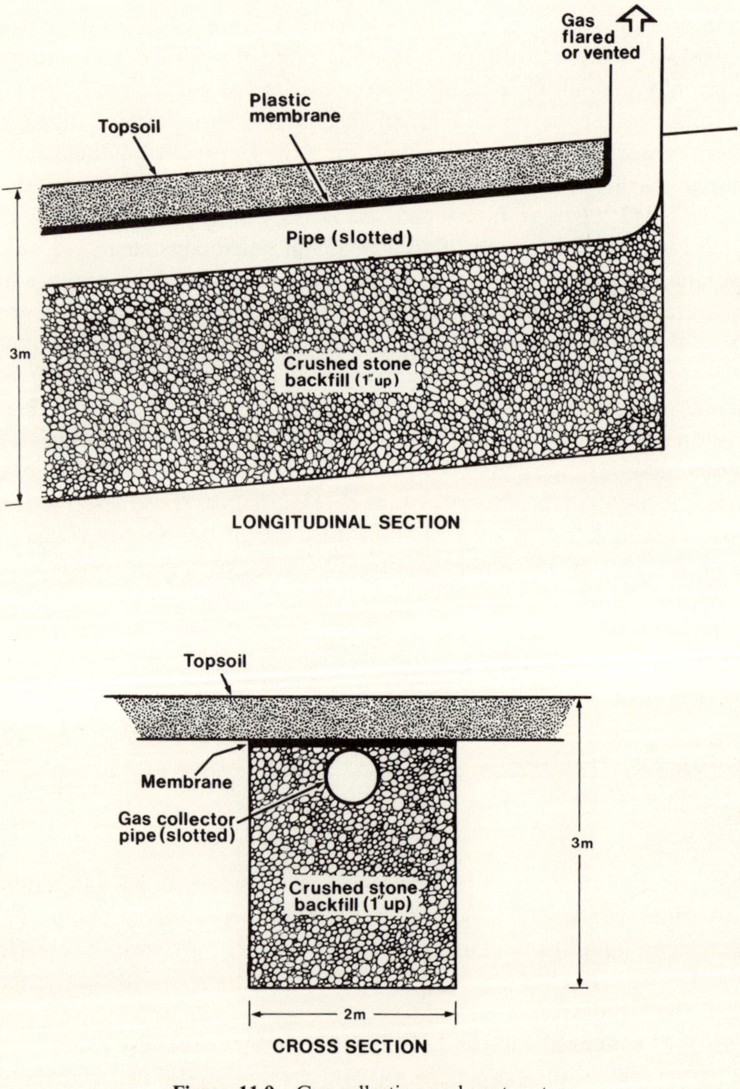

Figure 11.9 Gas collection and vent system

major problems of using the techniques on landfill is not merely the range of heterogeneous ground conditions but the physical obstructions that can be encountered e.g. steel drums, blocks of concrete, timber beams. These present fundamental problems for piling operations particularly when they are encountered at depth. Large voids can also present problems for slurry trenching and grouting.

11.5.4 *Vent trenches*

One of the most common off-site gas migration control measures used throughout the country is the vent trench. Basically this consists of a trench, usually dug outside of the tip material, and backfilled with crushed rock or gravel with, if necessary, a membrane on the outer face. Several of these systems have been installed with apparent success, some acting in a passive

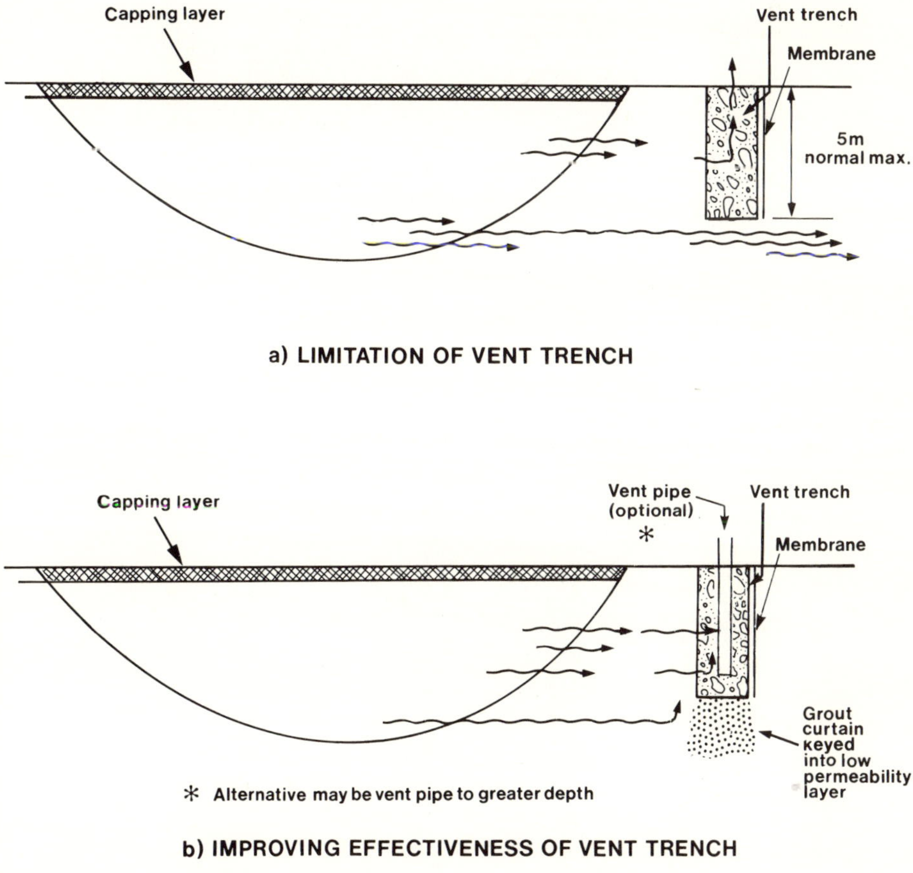

Figure 11.10 Improving vent trench function

way (i.e. relying on differential ground permeability and gas pressure and buoyancy for gas movement out of the trench), while some can act in an active way i.e, the system is used to collect migrating gas which is then drawn into a perforated pipe and pumped to a stack where it can be flared or merely vented, depending on the potential odour nuisance to the neighbourhood (Figure 11.9).

Depths of vent trenches are normally limited by the reach of a mechanical digger, i.e. to about 5 m. (unless the excavation is 'benched' in which case additional precautions are necessary for workers). This depth limitation can be compounded by trench instability and obstructions in landfill materials.

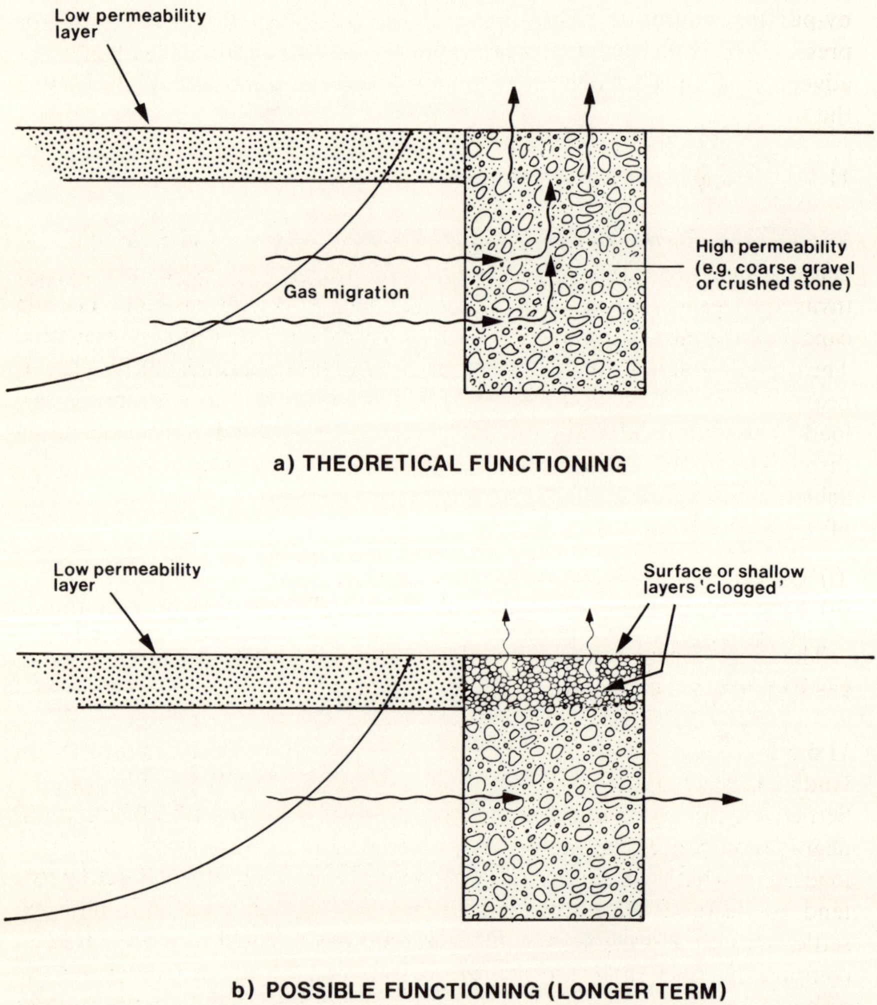

Figure 11.11 Theoretical and possible functioning of cut-off trench vent (without membrane)

Thus it may be necessary to introduce a further system below the vent trench (e.g. by means of a grout curtain) (Figure 11.10) to ensure that the gas does not pass beneath the trench.

Another factor to be considered in the use of cut-off vent trenches is the potential 'clogging up' of the surface, thus reducing their effective permeability (Figure 11.11). The implications of this can be obvious especially if a membrane is not used. Also in choosing a vent-trench gas control option, consideration must be given to the potential implications of the considerable quantities of gas emerging.

Setting a vent trench within a landfill is not likely to be fully effective in preventing migration from one part of the site to another, since it is likely to be by-passed underneath. Thus, while it would undoubtedly create a low pressure zone which would have an influence on gas migration, it could not be advocated as an effective measure unless it reached to, or close to, the base of the tip.

11.5.5 *Ground engineering factors*

11.5.5.1 *Introduction.* The implications of methane and other landfill gas emissions are not solely confined to health hazards. These gases are, almost invariably, generated in organic material which usually has poor bearing capacity in terms of supporting structures, roads and infrastructure generally. The organic materials may be ubiquitous or in layers (such as in the case of peat) and these must be either consolidated sufficiently to sustain the new loading conditions without undue settlement, or by-passed by means of piling through to suitable underlying strata. The engineering techniques necessary to achieve either of these effects can have very significant implications in terms of:

(i) Gas migration patterns and routes, and
(ii) Providing potential pathways for gas to enter buildings or structures.

Therefore, consideration of the effects is crucial to the proper assessments of gas hazards in the context of site development.

11.5.5.2 *Ground settlement.* One of the most critical engineering problems on landfill sites is that of ground settlement, particularly differential settlement. Settlement can occur due to the self-settlement, resulting from the natural degradation process of the landfill material, and/or through the increased loading on what is, in general, poorly compacted ground. Rates of settlement tend to decrease with time but predicting relevant time limit on such settlements is not possible, e.g. rates of decomposition will vary with material composition and environments within the site. These variations are the principal causes of differential settlement and are consistent with the random nature of tipping processes.

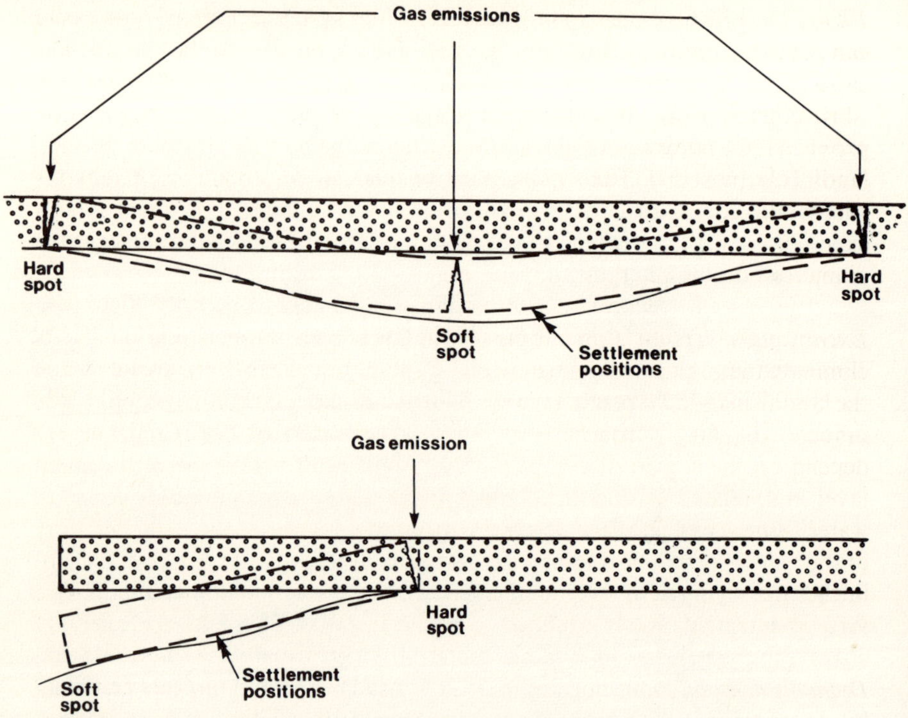

Figure 11.12 Effects of differential settlements on rigid slabs

The principal physical implications of differential settlement is the discontinuity usually caused to the overlying material. Thus a clay cover is likely to shear locally, while a concrete slab would either 'break its back' over the 'hard spot' or deflect as illustrated in Figure 11.12.

The simplified consequence of these settlements is that gas can penetrate what was previously a robust, high integrity barrier. Moreover, the volumes of gas emerging through such pathways can be much greater as they represent the emission levels from a much wider area, i.e. this may be the only, or most suitable, outlet.

11.5.5.3 *Ground improvement techniques.* The following are the techniques normally employed or considered for improving the structural characteristics of poor ground:

Piling
Excavate and replace
Dynamic consolidation
Vibroflotation
Preloading/surcharging
Grouting

Piling. The principal relevance of piling *vis à vis* landfill gas hazard is that piles can provide a pathway through which the gas can migrate to the surface. The capacity of this pathway can be increased over time as the landfill mass may 'shrink' away from the pile. (This can also present a serious engineering problem if the pile loses its lateral support.) In any event, driving piles through landfill can present considerable physical problems when obstructions such as drums and lumps of concrete are encountered. Similar difficulties can be experienced with bored piles, which are often favoured due to the large diameters attainable, thus allowing higher safe working loads.

Excavate and replace. Pursuing this technique to the limit would obviously eliminate the gas hazard, but there may be some relevance in removing part of the landfill mas and replacing it with a more suitable engineering material. The influence of this approach on either gas generation or gas migration will depend on the degree of compaction achieved both within the replacement layer and, related to this, the change to the underlying material in terms of water content and temperature. Notwithstanding either of these, the integrity of the new layer will be directly related to the degree of differential settlement in the underlying layers. Thus the scenario is similar to that of a rigid slab *vis à vis* integrity.

Dynamic consolidation. This technique consists of repeated impacts on the fill by dropping a large weight from a height which can be up to a (current) maximum of 20 tons and 20 m respectively. This process is carried out on a regular pattern with the chosen weight and height being governed by the depth of fill. Charles (5) suggests that for domestic refuse the likely maximum depth of compaction is 8 m. The impacting process is repeated at closer intervals to compact the shallower zones.

There are a number of potential consequences from using this technique. Gas pathways can be created along the sheared faces of the landfill material, i.e. the column of material under the dropped weight is compacted relative to the adjacent material and, hence, discontinuities occur; gas is forced out at high pressure and 'dry' refuse might have been pushed below the water table. Further relevant effects include: the displacement of leachate (or contaminated groundwater) into adjoining areas; the creation of ponding around the point of impact—'splashing' of such liquids can be hazardous to operatives.

Vibroflotation/vibroreplacement. Vibroflotation consists of inserting a vibratory poker into the ground and this effects a compaction of the surrounding material. The technique was developed for, and is best suited to, granular materials. In landfill and other sites however, the technique is modified by the introduction of crushed stone around the poker which then forms a column. From a practical point of view this technique can be difficult as obstacles within the fill can jam the vibroflot. The stability of these columns can be affected by reduced shear strength owing to 'slurries' lubricating the stone, and reduced lateral support due to degrading materials. The major implications of the technique for gas migration is that a 'chimney' is created for collecting and venting. It is also possible for the stone to be attacked by acids (e.g. limestone can be degraded and give off CO_2).

Preloading/surcharging. Preloading can produce an immediate settlement, and thus at least reduce the extent of longer-term settlements. This increased compaction is not likely to change the gas regime to a major degree, but would obviously increase the pressure and, as with dynamic consolidation, can increase the gas generation zone by pushing refuse into the saturated zone. The technique could also influence the location from which gas emerges since the cover to the site is, in effect, likely to be less permeable. On the other hand, any expulsion of water or a greatly reduced ground permeability can cause a deterioration in the methane generating environment.

Grouting. Permeation grouting in landfills to achieve engineering improvement has not been generally practised, largely due to problems in forming grouting holes and to the uncertainty of filling large voids. Were the technique used effectively, there would undoubtedly be a considerable improvement to the bearing capacity of the site. However, if successful, grouting could both change the water regime within the landfill (and so influence gas generation) and also increase pressure by reducing void space.

11.5.6 *Pumping*

The effect of pumping gas from (perforated) boreholes can be very significant and the effective zone of influence has been measured at up to 30 m. (6). This effectiveness does, however, depend on a number of factors such as

The suction pressure
The relative and absolute permeability of the local landfill mass
The permeability of the surface cover.

An effective balance in pumping efficiency can only be achieved by trial and error since there are many relevant variables, not least of which is the fact that excessive pumping can induce critically high oxygen levels within the tip and so kill the methane-producing bacteria. The provision of collector (perforated) pipes, or gravel/stone filled trenches associated with a well point, can greatly improve the efficiency of a pumping system since they not only intercept laterally migrating gases but also offer less resistance to pumping.

Gas pumping can also be achieved by setting perforated pipes into a horizontal layer of gravel placed between the landfill mass and a low permeability capping layer (Figure 11.13). This type of shallow system has the advantage of being relatively easy and cheap to install. However, because of

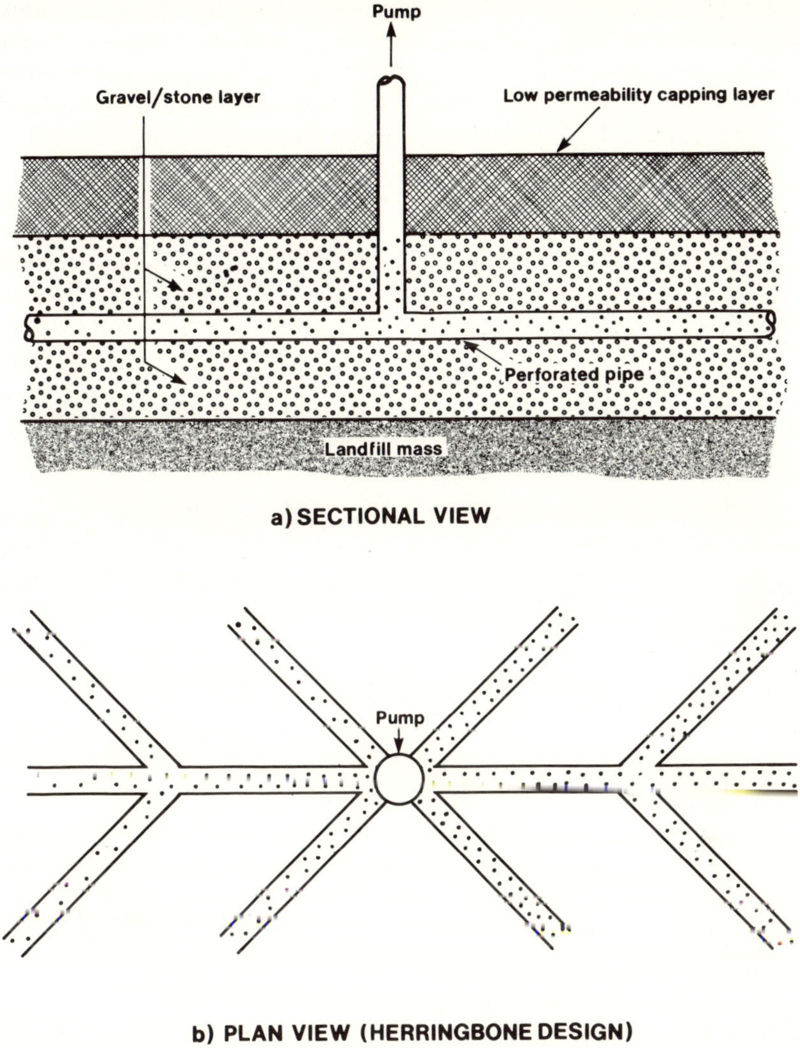

Figure 11.13 Shallow pumping system

differential settlement, the capping layer can be disrupted causing, on the one hand, an escape route for gas and, on the other hand, can result in air being sucked into the pumping system. A further significant disadvantage can be the mechanical stress placed on the collector pipes due to differential settlement.

Differential settlement can also create practical problems for the conventional well-point pumping or venting system as above-ground pipes may suffer through settlement of their supports. The loss of seal around the well pipe (where it passes through the capping layer) can also be critical since the effective pumping pressure within the landfill can be greatly reduced. Where a passive venting system is used, this seal is not, of course, essential.

11.6 Hazards in buildings

11.6.1 *Introduction*

The problems and risks associated with methane in buildings generally, and confined spaces in particular, have been given serious consideration long before problems from landfill gases became topical. Many guidelines and reports on the subject have been produced, particularly in the coal and mining industries. Since the principal danger from methane is that of explosion, precautions should be taken to prevent critical concentrations. These precautions vary from removing it by ventilation, diluting it with air, and/or monitoring the incidence of the gas. BS 6164 (*Safety in tunnelling in the construction industry*) (7) states that if the concentration cannot be kept consistently below 0.25% (approximately 1/20 the LEL value) in the general body of air, explosion protection will be required. 'Where a concentration of 1.25% is present, men, other than those essential for safety of other personnel, are to be withdrawn from all parts of the tunnel...'

The importance of CO_2 as a hazard must also be kept fully in mind. The effects of critical concentrations might be less spectacular but can be equally fatal. Apart from the early stages of decomposition, however, CO_2 is unlikely to exist as the sole hazard from landfill gas. Thus, efforts to abate methane hazards will invariably deal with CO_2.

11.6.2 *Explosions*

Because methane is a combustible gas, an explosion hazard can arise. When the gas is ignited, the following hazards could arise:

People are knocked over, shocked, injured, burned or killed
Buildings are burned
Buildings suffer structural damage.

The result of ignition is the rapid propagation of flame which raises the temperature of the gas and, without venting, the pressure will increase. Pressures as low as $15\,kN/m^2$ ($2\,lb/in^2$) are sufficient to cause structural damage to buildings. This force, added to the risk to human health caused by the flame, can result in very serious consequences.

If the methane concentration is in excess of 15% then, in theory, the gas will not take part in any explosion. However, it is inevitable that in a dynamic situation some portion of the air/gas mixture is in the critical 5–15% range. Once an explosion has been induced, mixing will occur and so the 'excess' methane will contribute to the explosion. It is, therefore, necessary to consider any level above the LEL as presenting an explosion hazard.

11.6.3 *Gas mixing and layering*

When two or more gases are thoroughly mixed, they cannot separate out into their individual buoyant layers (at least in the range of possible conditions under consideration). Therefore, if a gas mixture is introduced to a room, the new mixture (no matter how poorly mixed) cannot be explosive *if* the gas entering is less than the LEL, and if the room atmosphere is 'normal' air. Thus, it is reasonable to conclude that if methane is introduced at less than the LEL limit, that limit cannot be exceeded even in a confined space.

A buoyant gas entering a room at a low position will tend to mix well with the air as it rises, especially if it enters away from the walls. Conversely, if the gas enters through a wall which is smooth, it will tend to 'adhere' to it without good mixing and this may form a stable stratified layer at the ceiling. The volume of gas entering, the height of the room and the quantity of ventilating air will all affect the gas-mixing potential.

In simple practical terms, the maximum ventilation rate for a room, into which methane leaks, must be up to 15 times the leakage rate if the methane concentration in the room is to be kept below 5%, i.e. the maximum likely methane concentration of 75% divided by 15. However, for this rate to be effective, it would be necessary for the 'fresh air' and methane to be mixed completely. In a 'normally' used room where there are windows and doors, it is reasonable to expect considerable mixing, with the movement of occupants and the opening of windows or doors. In many modern houses, however, where natural ventilation is low due to, for example, double glazing, then a hazardous situation can arise more readily. Other classic examples of such occurrences are 'broom' cupboards, rooms which are used infrequently, or rooms that have not been used for a protracted period, e.g. over holiday time. Such a circumstance occurred at a Christchurch recreation centre in 1979 (8) and, more recently, the effects were catastrophic at the Abbeystead pumping station (9).

11.6.4 *Acceptable size of gas accumulations*

No firm conclusions have been reached on the acceptable size of a gas accumulation. Since the hazards from ignition relate both to people and buildings, a different size of acceptable accumulation would arise for each category of hazard. Many factors can influence the pressure produced from gas ignition. These include:

Volume
Gas properties

Concentration
Degree of pressure relief or venting
Possibility of pressure piling (i.e. successive pressure increases with successive explosions in connecting rooms).

Since it is not practical to monitor the volume of gas present in any particular location, it is necessary to consider the gas concentration as the sole determinant of hazard and no gas concentration, of whatever volume, should be allowed to exceed the specified value.

The questions of safety standards must, to a degree, be related to the facility by which those standards can be monitored. Taking an extreme view, methane concentrations of less than the LEL are safe, i.e. up to about 5%. But such a standard (were it used) would beg the following questions, for example:

How accurate is the monitoring device?
Where is the monitor positioned?
Is the concentration likely to increase before, say, an automatic ventilation system is brought into operation?
Has the ventilation system a 'fail safe'?
Has the system been maintained?

It is, therefore, necessary to have a large margin for error which takes account of these and any other relevant questions.

11.7 Effects on plants

11.7.1 *Introduction*

Strong correlations appear to exist between high concentrations of landfill gas in soil and vegetation die-back. On the other hand, however, the apparent ease with which some plant species grow on landfill sites cannot always be interpreted as plant tolerance, but might merely reflect the fact that gas is not reaching the plant roots in that area, or the gas composition is not critical. Similarly, where die-back does occur it cannot be simply assumed to be due to gas effects rather than to other effects such as phytotoxicity (from ground contamination by chemicals), poor soil structure, nutrient deficiency, drainage problems, or any combination of these. It must be borne in mind that, notwithstanding any gas effects, problems can be frequently encountered in revegetation of disturbed land.

Many researchers e.g. (10, 15) have observed the general similarities between landfill gas effects and the effects due to waterlogged conditions. This similarity is not surprising as both circumstances may exhibit, *inter alia*, the following conditions:

Anaerobic conditions due to

(a) gas presence
(b) water saturation.
Low pH
(a) acid surface or groundwater
(b) acidity due to CO_2 plus H_2O forming H_2CO_3 or peaty, organic conditions.

Hitherto, the general simplified explanation for vegetation die back has been that methane (which is frequently considered the only relevant gas) causes oxygen deficiency in the plant root zone without actually being toxic. While this explanation has, undoubtedly, considerable validity at high volume emissions (as opposed to high concentrations) it is a gross over-simplification of the potential influence of landfill gas on the already complex growth medium and growth process (see Figure 11.14).

11.7.2 *Potential toxic mechanisms*

11.7.2.1 *Introduction.* As indicated earlier, landfill gas can consist of a very wide range of gases, the majority of which occur in trace concentrations only. Considering the chemical and physical complexity of site conditions in the plant root zones of reclaimed sites, it is not surprising that research work has concentrated on assessing the possible effects of the two principal gases, methane and carbon dioxide. In the early stages of refuse decomposition, landfill gases can sometimes consist almost totally of CO_2, with methane production gradually increasing and normally overtaking the CO_2 con-

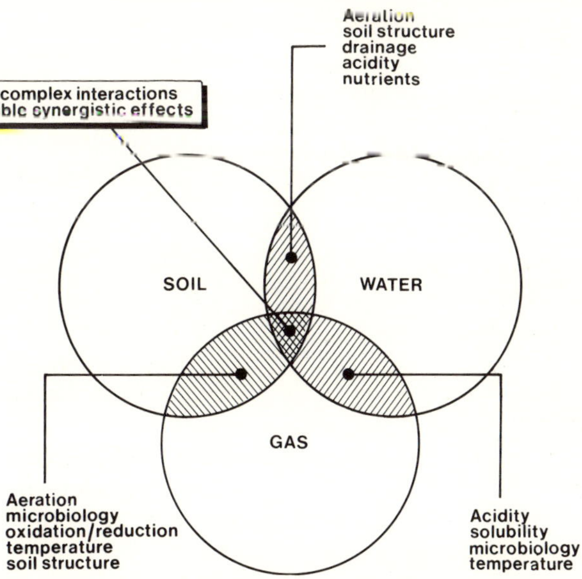

Figure 11.14 Influences on plant growth

centrations. Thus CO_2 can be critically present from the earliest stages, and this gas is widely accepted to be toxic to plants if present in excess.

11.7.2.2 *Oxidation of methane.* Since, under anaerobic conditions in particular, methane is a normal end product of the microbial fermentation of carbon-containing materials, minute traces of methane occur naturally in normal soil (ambient concentrations of methane in air is usually about 2 ppm) (10). Most soils contain micro-organisms capable of oxidizing methane *if* air is present, and Hoeks (11) and Pankhurst (12) report that, in the presence of natural gas, these bacteria multiply rapidly, especially when temperatures are elevated. Under some circumstances, such as low emission rates, these bacteria could oxidize virtually all methane present resulting in no release to the atmosphere. The process of oxidation is as follows:

$$CH_4 + 2O_2 \rightarrow CO_2 + 2H_2O$$

Thus, in soils in which methane oxidation is occurring, two particularly relevant effects can occur, namely:

(i) Oxygen concentration can be depleted
(ii) Carbon dioxide is produced.

Another important feature of the oxidation process is that it is an exothermic reaction (i.e. it produces heat).

It is clear that in many circumstances the concentrations of CO_2 could be critically elevated in the plant root area although measurements at depth might indicate otherwise. Where the oxygen concentration is low, the conversion of methane to CO_2 is not complete and intermediate products such as methanol, formaldehyde and formic acid are produced (13, 14). These can remain in the soil and exhibit direct toxicity to plants; formaldehyde, for example, is used as a fumigant in glasshouse stabilization programmes. Notwithstanding this, toxic minor components can also be present.

11.7.2.3 *Direct effects of CO_2 and CH_2.* Current scientific evidence is that CO_2 is toxic to vegetation (15) while the principal effect of methane is to produce an oxygen deficient environment in the root zone. A combination of both gases therefore (which is normal) at least causes a depletion of oxygen and is toxic. Whether or not these two gases act synergistically together with other minor gases is not known, but it is reasonable to surmise that the effects are likely to be at least additive, particularly at critical plant growth stages such as germination.

Acknowledgement

A large proportion of the information presented in this chapter was collected in the course of a review project funded by the Department of the

Environment's Central Directorate on Environmental Pollution (CDEP) and carried out by the author. The views expressed are those of the author and they do not necessarily coincide with those of DoE. The DoE's permission to publish this chapter is gratefully appreciated.

References

1. Rees, J.F. (1981) Landfill gas: major factors affecting methane production in landfills. Landfill Gas Symposium, (Paper 2), Harwell.
2. Peckson, G.N. (1985) Methane and the development of derelict land. London Environmental Supplement No. 13. ISSN 0264–7524.
3. Health and Safety Executive (1984) Guidance Note EH40: Occupational exposure limits 1984.
4. Naylor, J.A., Rowlands, C.D., Young, C.P., and Barber, C. (1978) The Investigation of Landfill Sites. Technical Report TR91, Water Research Centre, Stevenage.
5. Charles, J.A. and Burland, J.B. (1982) Geotechnical considerations in the design of foundations for buildings on deep deposits of waste materials. *The Structural Eng.*, **60**A (1).
6. Baker, W. (1985) City of Birmingham Engineer's Department. Personal communication.
7. British Standards Institution (1982) BS 6164: 1982. Code of Practice for safety in tunnelling in the construction industry.
8. County Surveyors Society (1982) Gas generation from landfill sites. Special Activity Group No. 7, London.
9. Health and Safety Executive (1985) *The Abbeystead Explosion*. HMSO, London, ISBN 0 11 8837958.
10. Pankhurst, E.S. (1977) The effects of natural gas on trees and other vegetation. Welsh Soils Discussion Group Report 14, 116–130.
11. Hoeks, J. (1972) Changes in composition of soil air near leaks in natural gas mains. *Soil Science*, **113** (i): 46–54.
12. Pankhurst, E.S. (1978) Bacteria, barnacles and biomass. Communication 1070, 44th Autumn Meeting. The Institution of Gas Engineers. 21/22 Nov. 1978, London.
13. Brown, L.R., Stravinsky, R.I., and McClesky, C.S. (1964). The isolation and characterisation of *Methanomenes* methane oxidation. *Can. J. Microb.*, **10**, 791.
14. Nyns, E.J. and Wiaux, A.L. (1960) Biologie des Hydrocarbures. *Agriculture*, **17**, 3–56.
15. Arthur, J.J., Leone, I.A., and Flower, F.B. (1981) Flooding and Landfill Gas Effects on Red and Sugar Maples. *J. Environ. Qual.* **10** (4), 431–433.

Index